"*Dolphin Chronicles* gives a true and lively report of research into the intelligence and behavior of a large-brained nonhuman animal species. Emphasizing communication, *Dolphin Chronicles*'s message will have uncommon appeal for today's networking generation."

—Victor B. Scheffer, former U.S. government marine biologist and chairman of the U.S. Marine Mammal Commission, and author of *The Year of the Whale* and *A Natural History of Marine Mammals*

"Fresh, beautiful, and elegantly scientific, *Dolphin Chronicles* gives us a cascade of insights into our seagoing mammalian kin. Carol Howard's sensitive, dedicated research has produced a glorious book, a must for those who love animals and for those who love the sea."

—Elizabeth Marshall Thomas author of *The Hidden Life of Dogs*

Dolphin
CHRONICLES

A fascinating, moving tale of one woman's quest
to understand—and communicate with—
the sea's most mysterious creatures

Carol J. Howard

BANTAM BOOKS
New York   Toronto   London
Sydney   Auckland

DOLPHIN CHRONICLES
A Bantam Book

Library of Congress Cataloging-in-Publication Data

Howard, Carol J , 1949–
Dolphin chronicles : the two worlds of Echo and Misha / Carol J Howard
p   cm.
ISBN 0-553-37778-7
1. Atlantic bottlenosed dolphin   2 Atlantic bottlenosed dolphin—Behavior
3 Atlantic bottlenosed dolphin—Training   4 Atlantic bottlenosed dolphin—
Research—United States   I Title
QL737 C432H685   1995
599 5'3—dc20   95-16669
                                                                                           CIP

Published simultaneously in the United States and Canada

Bantam Books are published by Bantam Books, a division of Bantam Doubleday Dell
Publishing Group, Inc Its trademark, consisting of the words "Bantam Books" and the
portrayal of a rooster, is Registered in U S Patent and Trademark Office and in other
countries Marca Registrada Bantam Books, 1540 Broadway, New York, New York
10036

PRINTED IN THE UNITED STATES OF AMERICA

BVG 01

To my mother, Lorraine Leavitt Howard,
and father, Robert Bruce Howard,
who taught me to love the water

The dolphins lord it greatly among the herds of the sea, pluming themselves eminently on their valiance and beauty and swift speed in the water; for like an arrow they fly through the sea, and fiery and keen is the light they flash from their eyes.

—Oppian

Contents

Foreword

As I looked forward to retirement after almost four decades of research work on dolphins, a most curious and poignant paradox became evident to me. It seemed that much of what we had learned in that span of years might never be expanded upon. Soon, if public opinion about research with captive dolphins followed the dictates of those espousing animal rights, we might have no chance to learn more, even though dozens of tantalizing avenues, pregnant with probable discovery about dolphin lives and capabilities, were just then opening.

A small, vocal, and insistent group of people were calling into question the capture, holding, and study of dolphins. Although at first they seemed to be a far-out minority, their views were increasingly being melded into the mainstream. The way these things work, in a generation's time, what had been merely clamor at first was becoming the prevailing view.

What hit me hardest about this transition was the insistence of the animal-rights community that we cut off our most intimate and direct contact with dolphins; that we live with the myths many people had spun in an impenetrable web around dolphins instead of continuing to seek out true things. Viscerally, deep down, I have come to rail against this curious propensity of humans. As a scientist I do not accept that such myths give viable directions about how we should live in this world. Too many excesses and misdirections of our species have sprung from this tendency to build castles in our imaginations and then live by them. Over and over again in my career, the true things that my colleagues and I have uncovered have come to seem more wonderful by far than anything we could have made up. We had by our science forged this connection to the dolphins, and to nature more generally. This bridge had come to seem to me very precious indeed.

At the same time I also understood and resonated with the emotional underpinnings of the animal-rights movement. The more we have learned about the lives of such wild animals, especially the birds, wolves, elephants, chimpanzees, hippopotamuses, and dolphins of this world—and even those animals more remote still—the more we find we have in common. No wonder our care for them is awakening!

My house is full of cats and dogs, parrots, peafowl stomping on the roof, yelling out their loves for each other, chickens all running free, digging deep in the dust to make their baths down in the azalea garden. They are no inconsequential part of our lives here at my farmlet of *El Encinal,* as I call it, named for the oak grove outside the windows. It's a little harder to do this sort of thing with dolphins, their housekeeping needs being what they are, but I've managed that too. What has mattered most in all these connections, dolphins included, were the modest things between me and them that I could see, hear, and feel—and sense on the emotional plane. They are different from us, yes, but all of them are sentient sharers of the Earth with the rest of us. I could not have come to understand that fact locked in a thermostatically controlled flat in Brooklyn with only an African violet to keep me company. The bridge between us is crucial, it came to seem.

But when you build a mythic connection, it is easy to shove it aside because deep down you know it is merely a myth. Ultimately, the true things we have learned about other species are what matter; they are the connections that one can depend upon. Such true things can cause a person to step back and say: "No, I must not do that! It will hurt my friend." I most deeply want real animals at the other end of my communications; I want to feel those links of touch, smell, sharing of food, dividing of space, and the even more subtle building bridges of understanding between us.

So, as my last scientific effort before retirement, I decided to see if there was a compromise possible between the demands of the animal-rights community and dolphin scientists. Could we scientists learn to "borrow" a cadre of dolphins from nature, bring them into our laboratory, learn from them, and release them again to live out the rest of their wild lives? I called it "The Dolphin Science Sabbatical." My hope was that the "dolphin connection" could remain unbroken and that we two species could continue to build our relationship based upon truth, not myth. This proved to be about as difficult a venture as I ever undertook.

Just before I began this effort, into my intellectual orbit came a quite remarkable young woman—Carol Howard, the author of this book. She was not a biologist, which is how I am classified, but rather a psychologist, and a skilled writer about sciences for more public audiences. She subsequently became a doctoral student in my laboratory and carried out the scientific studies of two dolphins we captured for the world's first "dolphin science sabbatical."

In spite of the differences in our knowledge and training, I came to regard Carol as possessing a very special, incisive talent. Time and time again, in her quiet, measured way, Carol would cut through the verbiage of our discussions to reveal the exciting, central insights that were to be had. She sees with remarkable clarity of mind and writes with an unvarnished integrity—both things evident over and over in this book. Beyond that lies Carol's deep emotional connection to the world and all that is in it.

Did the "science sabbatical" idea succeed? Can we think of borrowing dolphins for a while to learn from them?

No. In the last analysis I finally discarded the idea of the dolphin sabbatical, at least as we performed it, as a viable way of providing a connection between our two species. It proved far too difficult a traverse for naive dolphins to take over and over again on a rather short time schedule. Misha and Echo were pushed too hard by the transition we imposed upon them. The problem was not so much the capture, or the transportation across the country by airplane. It was, rather, the difficulty of telling a fresh-caught wild dolphin that we wanted to learn from it. In retrospect, we now see that those two dolphin teenagers began their "sabbatical" not knowing that we were even asking questions of them. Repeatedly we humans confronted them with abrupt and frustrating shifts of behavior, which we thought were clear commands, but which instead presented them with unfathomable double binds. They simply didn't know we were using "yes-no" as a bridge between them and us. In the end, however much I wanted to open the channel between us and then explore a little, this was, I concluded, not the way it should be done.

No, if the "dolphin connection" is to be kept open, a certain number of dolphins must remain in captivity, must serve as ambassadors for their species. We must allow all the time a naive dolphin needs to develop gradually what the comparative psychologists call "a set for learning"—that magic time when the lightbulb goes on over the animal's head and it knows that it is being asked questions. Then it can become a game—one that both dolphin and human can love to play, one that enriches the lives of both participants.

Kenneth S. Norris
El Encinal, Bonny Doon,
California
14 January 1994

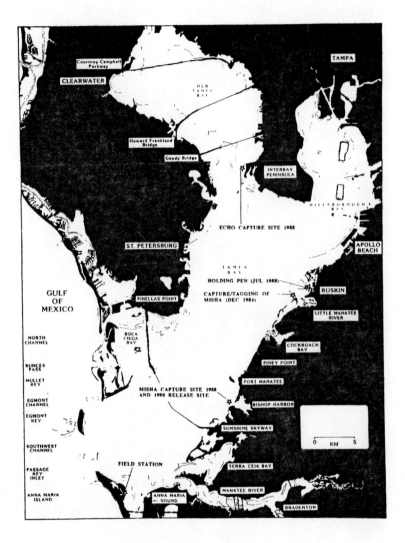

Introduction

I've never seen a UFO, much less an extraterrestrial. Nor have I ever met any of those myriad people the supermarket tabloids claim have been abducted by such visitors from outer space. But I do know a couple of creatures who were indeed carried off by alien beings, studied intensively for a time, and ultimately returned to their own world. I know because I'm one of the aliens.

No, I don't hail from some distant galaxy. I'm a red-blooded, air-breathing, Earth-born mammal. So are the creatures I helped carry off—two Atlantic bottlenose dolphins, whom we came to know as Echo and Misha. I was part of the team of scientists, students, and volunteers who caught these two dolphins in Tampa Bay, Florida, in July of 1988 and flew them to Long Marine Laboratory in Santa Cruz, California. We spent the next two years observing and recording their sounds, their movement patterns, and their social interactions. In October 1990 we took the pair back to their home waters and released them. We've been carrying on a long-term follow-up study ever since, checking in on them regularly to see how they are readjusting to the rigors of life in the wild.

I've been intimately involved in all aspects of this project—as a member of both the capture and release teams, as one of the dolphins' principal trainers, and as a graduate student whose doctoral dissertation is based on the research we conducted with them. Echo and Misha are a lot more than mere research subjects to me. I didn't just collect data from them; I fed them, played with them, talked to them, stroked them, swam with them, spent hours just watching them, worried when they were sick or not eating properly, and regaled friends and acquaintances with pictures and tales of their exploits. I cried when they left. I still miss them enormously.

This approach of bringing animals into captivity for a relatively short period of time and then releasing them again—what we referred to as a "dolphin science sabbatical"—was a major thrust of our research with Echo and Misha. We were hoping to find a "best of both worlds" solution to the question of keeping dolphins in captivity, a way to keep open the channels of communication between humans and dolphins without making captivity a life sentence for the animals.

The "sabbatical" didn't turn out as well as we had hoped in some regards, but far better than we could have imagined in others. It probably isn't the way to go for either research or show facilities. For one thing, it's too expensive. More important, it's too stressful for the animals—just about the time they start getting the hang of life with humans, you turn around and ask them to go wild again. We did, however, learn a great deal about how to ease that transition between captivity and the wild, how to tell which dolphins might be good candidates for release, where to release them, and how best to ensure their successful reacclimation. Some of these guidelines have already been applied to subsequent releases of captive dolphins and of injured wild dolphins that were rehabilitated and then returned.

Our findings are especially important these days when there is considerable pressure to "Free Willy" and all his dolphin kin. But the movie doesn't tell you how Willy was doing a month or a year or three years after he was freed. Everyone simply assumes that he readily readjusted to life in the wild. Not necessarily a valid assumption. The tale isn't quite so heartwarming if Willy starved to death because he didn't know how to feed himself. Or if he wasn't accepted into a killer whale pod in the area and lived his life as a social outcast, perhaps even harassed or attacked by the others—not an easy fate for a creature as sociable as an orca. Or, worst of all, if he carried some disease for which the local animals had no immunity and, in exercising his freedom, managed to decimate the native population.

Ours wasn't the first release of captive dolphins, but it was undoubtedly the best planned and the most thoroughly followed up. Other dolphins besides Echo and Misha have been released—some to considerable press play—but nobody knows for sure what happened to those animals once they swam off into the blue. We know what happened after Echo and Misha returned home.

This book tells Echo and Misha's story, albeit as seen through "alien," human eyes. It isn't just a tale of two dolphins, but a tale of two worlds as well. With Echo and Misha we've had the unprecedented opportunity to get to know the same individual dolphins both in captivity and in the wild, in our world and in theirs.

Though we're increasingly able to study dolphins in the field, we humans remain limited in our ability to make ourselves at home in their aquatic habitat. Even in a boat it's difficult to track dolphins at sea for any length of time, and then we generally see only the surface of their lives—most of their activity takes place under water, beyond our purview.

Working with dolphins in captivity gives us a glimpse beneath the surface; it allows us to observe and interact with them in ways that simply wouldn't be possible otherwise. But we always have to bear in mind that dolphins in a tank may not necessarily act the same as free-swimming animals. With Echo and Misha, for the first time, we've been able to bring together both views of the very same dolphins, to come to know them first up close, as individuals, and then to place them in a larger perspective, as members of the dolphin community at sea.

Along with chronicling the two lives of Echo and Misha—their life with humans and their subsequent return to the wild life—I've also tried to characterize their perception and communication abilities, their senses and sensibilities, so the reader may get some feel for the dolphins' perspective on all this. My own research has focused on dolphin echolocation, their exquisitely sensitive and sophisticated sonarlike capability. Not only do dolphins live in a world very different from our own, they also perceive that world through a sense we can't fully imagine, one that uses sound, somewhat as we use sight, to give them a picture of their surroundings. We can't begin to understand the dolphin world, or the dolphin mind, without a better understanding of this extraordinary sensory system.

In this research, too, we were looking for a "best of both worlds" approach. Echolocation is nigh onto impossible to study in the wild. We wanted to take a step toward studying echolocation under somewhat more natural conditions, working with free-swimming animals rather than dolphins that were physically restrained, as has been the case in most echolocation research. We looked at what dolphins actually *do* as they echolocate, how they move and how these movement patterns relate to the patterns of sound that they make. It's something like looking at the facial expressions, hand gestures, and the like in human speech—movements that convey so much of the meaning of what we say, beyond just the words themselves.

This book also serves, in part, as my long answer to a question I've been asked many times since I first started working with dolphins: "So how smart are they *really*?" My short answer is simply "I don't know." I can cite views ranging from John Lilly's contention that whales and dolphins are "more intelligent than any man or woman" to those that would grant them all the intellect of a hedgehog and the social graces of a cow. But I think the question is wrong. How do you judge an alien intelligence? How smart are we, as measured in dolphin terms? I know if you set me out at sea amid a dolphin school, I'd flunk out in no time.

What we think of as intelligence relates, I believe, to the degree of complexity and flexibility in an individual's behavior. My goal hasn't been so much to place dolphins on some cosmic hierarchy of intellectual abilities as to introduce Echo and Misha as unique, compelling individuals who behave in complex and flexible ways. I haven't tried to measure dolphin intelligence but rather to illustrate it—to show that intelligence at work in Echo and Misha's behavior, both in captivity and in the wild. I have tried to convey both the familiarity and the profound "alienness" of that intelligence.

In many respects dolphins are about as different from us as any mammal can be. In evolutionary terms our paths diverged some sixty million years ago when their ancestors, an early form of ungulate (yes, it's true, they *are* more closely related to cows than to us), left the land and ventured back into the ocean. Their bodies, their brains, their sensory systems, all changed dramatically over time in response to that new environment—an environment so alien to us that we must approach it as we would another planet, wearing air tanks and an array of special gear if we are to spend any time there.

So why do we see some reflection of ourselves in these very alien creatures? Why do we look at that built-in curve of the mouth and see a beneficent smile? Why are we so eager to see a humanlike intelligence in that fishlike form? *Humans of the Sea,* John Lilly subtitled one of his books; Belgian oceanographer Robert Stenuit called his *The Dolphin, Cousin to Man.* (I don't suppose *The Dolphin, Cousin to Cows* would sell many copies. Actually, their closest living relative, outside the whale clan, is probably the hippopotamus.) What is it about dolphins that makes us want to claim them as our long-lost kin? Or what is it about us?

I suspect part of the answer may lie in what anthropologist-essayist Loren Eiseley referred to as "the long loneliness"—we humans have no one to talk to but ourselves. What many people really want to know when they ask "How smart are dolphins?" is "Can we talk with them?"

I'll confess, that's the question that first drew me into this business. Ever since I can remember, I've dreamed of talking with the animals. My childhood readings consisted mostly of animal stories and fantasy books. I spent much of my youth in Oz and Narnia, Wonderland and Middle Earth, where people conversed readily with lions, walruses, and trees. Much later I heard about Dr. John Lilly's work with dolphins. He claimed that dolphins communicate via a language equivalent to ours and predicted we would soon be able to decipher this "Delphinese." I thought at last I'd found the real thing.

Unfortunately, Dr. Lilly's extravagant claims regarding dolphins' superior intellect, sophisticated linguistic abilities, and advanced ethics remain largely unsubstantiated, and he has long since lost most of his scientific credibility. (His LSD studies, on dolphins as well as on himself, didn't help matters any.) But the issues he raised—intelligence, communication, ethical considerations regarding our treatment of dolphins—are still the ones nearest to my heart, the ones that inspired this book.

John Lilly may be the name most frequently associated with dolphins by the general public; but I'll bet if you asked the people who actually work with marine mammals to name the grand old man of the field, the response you'd hear most often by far would be Dr. Kenneth S. Norris. Truly one of the pioneers of modern dolphin and whale research, he's been observing their behavior, both in the field and in the lab, for more than thirty years. He's been a professor, an amazingly prolific researcher and writer, and a lobbyist on behalf of whales and dolphins, helping to bring about great changes in public attitudes and policies toward these animals. He has written two popular books about dolphins—*The Porpoise Watcher* and *Dolphin Days: The Life and Times of Spinner Dolphins*—as well as any number of magazine articles, including two in *National Geographic.*

I have been most fortunate to have Ken Norris as my mentor and graduate adviser. He combines a scientific outlook with a lively imagination and an appreciation of the unexpected. He is a naturalist par excellence, with a gift for observing the animals themselves and letting them—their behavior, the way they are constructed—tell him what they are about. With Ken's assistance my idea of "the real thing" has shifted. I've become less concerned with the question "Can we talk with dolphins?" and more concerned with questions of who they are and how they lead their lives.

In a sense this book follows not just Echo's and Misha's sojourns but my own journey as well, my efforts to come to know dolphins as living, breathing animals, not just as a source of fantasy fulfillment. Many people seem to find it funny that I became interested in dolphins while living in such a supposedly landlocked state as Minnesota. What most don't realize is that Minnesota—the "Land of 10,000 Lakes," after all—has more shoreline than Florida, California, and Hawaii combined. (The only problem is, all the water freezes solid for half the year.) I grew up loving water, spending most of the summer near, on, or in a lake. In the water I felt more graceful, less self-conscious than I did on land. I especially loved swimming under water and only wished I could do it better. My idols back then were the river otters, who swam with such grace and agility, who seemed to have a high time while they were at it, too. Then I saw dolphins. I was nearly thirty years old before I first encountered real, live dolphins at the newly opened Minnesota Zoo. My life would never be the same. Something in their eyes, something in the way they moved, got to me.

Dolphins seemed to bring together nearly all my major passions at once—a love of the water and of animals, and a compelling need to understand the processes of communication and language. Human language has been a major source of both frustration and fascination throughout my life. Speech has never come easily to me, and for a long time I was sure there must be some trick to it I didn't know, some secret I was missing. Virtually all my schooling and work experience has, in one way or another, concerned the questions of language—what it is, why we have it, how it works, and where it fails.

My career has followed a quite circuitous path, cutting through several fields along the way, in my search for answers to these questions. As I earned my bachelor's degree in psychology back in the 1970s, my central interest lay with efforts, just then beginning, to teach human sign language or artificial language systems to chimpanzees. I went on to do three years of graduate study in speech pathology, dealing with people with a variety of speech and language impairments. I worked as an instructor in a program for mentally retarded adults with behavior problems; in all cases at least part of the problem stemmed from communication difficulties. I switched course somewhat after that and worked four years in a university English department, where I gained a greater appreciation for the power and beauty of language. That made me decide to try my hand as a writer, so I enrolled in the graduate program in science writing at the University of California at Santa Cruz.

Part of the appeal in moving to Santa Cruz (besides water that doesn't freeze over) was that I knew that's where Ken Norris taught. I figured I just might be in the right place at the right time. I was, as luck would have it. Ken was about to begin a research project in the newly completed dolphin facility at Long Marine Lab and needed a graduate student to work with him on it. Fortunately for me, he was intrigued by my rather eclectic background and agreed to take me on. I had my chance at last.

I've found this business of learning about dolphins to be always fascinating but not always easy, in part because I lacked the usual training in biology at the outset, and in part because I've repeatedly had to let go of preconceived notions of who these creatures are—or who I'd like them to be. One problem with all the modern-day hoopla about dolphins is that it's too easy to get caught up in the mythology and fail to see the animals themselves. As Ken Norris says, "Myths are fun, but I see a marvelous animal that's still very much of a mystery."

My aim in this book has been to present a view of dolphins that is well grounded in science but within a framework that allows the heart and soul of these creatures to show through as well. I have tried to portray Echo and Misha as multifaceted individuals. They are down-to-earth flesh-and-blood animals who nonetheless seem to inspire some sense of the mythic, of the mystical in us. They are strangely familiar, and yet so alien.

I believe we stand to learn much by getting to know these alien beings from our own planet. Psychologist C. G. Jung once argued that we humans are incapable of truly understanding ourselves because we have nothing with which to compare ourselves and therefore have no criteria for self-judgment. "The possibility of comparison and hence of self-knowledge," he concluded, "would arise only if [we] could establish relations with quasi-human mammals inhabiting other stars."

Dolphins are mammals certainly, with at least some "quasi-human" characteristics, and they inhabit a genuinely alien world, if not other stars. Even if we never achieve the in-depth philosophical discussions with them I once dreamed of, they may still provide us some perspective on ourselves, some broader basis for self-knowledge. We're in no position to worry about establishing relations with extraterrestrials, should there be any, when we're still so woefully ignorant of our relationship to our many "coterrestrials."

1

Destiny Drive

*S*huffle your feet, just shuffle your feet," Larry called down at us, his southern drawl as thick as the Florida summer's air. Perched atop the sighting tower of his fishing boat, he shouted out instructions like a square-dance caller, directing the rest of us in the steps of "the stingray shuffle." Most of the members of our twenty-two-person crew were wading through waist-deep water just offshore from Cockroach Bay, along the eastern edge of Tampa Bay. Such shallow, sandy areas tend to be a favorite haunt of stingrays. They like to settle on the bottom and flutter their way in under the sand, rendering themselves nearly invisible. In the murky bay waters it's impossible to see your own feet, let alone what they might be landing on. That was the problem.

Stingrays are placid enough creatures usually, and beautiful, almost hypnotic to watch as they glide along with graceful undulations of their thin, flat bodies, like waterborne flying carpets.

Stepping on one is another matter. The ray's long, whiplike tail is armed with a formidable multibarbed stinger—far worse than any fishhook—near its base. Not content merely to impale an intruder and leave it at that, a disturbed stingray will thrash about once it has implanted its stinger, further ripping the victim's flesh with each twist. Sometimes the two-inch stinger breaks off in the tussle and, with its row of tiny barbs down each side, will work its way in deeper and deeper if left untreated, like a porcupine quill.

The idea behind the stingray shuffle is to kick sand out ahead of you with each step to alert any stingrays in the neighborhood of your approach. The rays will simply swim off peaceably if they know you're coming, at least in theory. The method isn't infallible, however: one of the volunteers in our crew had twice before been stung by rays, requiring medical attention each time. So we shuffled our way cautiously through the water toward the net that encircled a small group of dolphins. That's what we were after—dolphins, not stingrays.

We were there in Tampa Bay to catch two Atlantic bottlenose dolphins. Ideally, we wanted a pair of young males, roughly four to seven years of age—old enough to have been away from their mothers awhile but still several years from sexual maturity. That was a key element in our plan for a "dolphin science sabbatical," in which we would essentially "borrow" a couple of years of the dolphins' lives, bringing them into captivity for two to three years of research and then returning them. We sought two young males with the idea in mind of ultimate release. Among the wild bottlenose dolphins in nearby Sarasota Bay, juvenile males are known to form close associations with one or two other young males, bonds that may last for many years, possibly even for the rest of their lives. If our two animals were to form such a bond, we figured, they might readjust more readily to life in the wild.

Although captive dolphins had been released on several previous occasions, what we had in mind was unique. For one thing, this was the first time such a release had been planned from the outset, even before capture. We also intended to release the dolphins back into their home waters, in the same area where we caught them. Furthermore, we planned to carry out a detailed, systematic, long-term follow-up study of the results—another first—and in fact had already begun studying the resident dolphin communities in Tampa Bay.

The dolphin sabbatical was originally the brainchild of Dr. Kenneth Norris, a world-renowned expert on whales and dolphins. Ken was the scientist officially in charge, the principal investigator on the current project—or the "Grand Pooh-bah," as we tended to call him when he wasn't around. Not that there is anything pompous about him. Far from it. Bearded, balding, given to wearing flannel shirts and baseball caps and driving a pickup truck full of animal feed of various sorts, he tends to look more like a prospector than some ivory-tower academic. The title was intended as one of affectionate veneration.

This was to be Ken's last big research venture before he retired. Its aims were twofold: one, a laboratory study of echolocation, dolphins' biosonar; and two, a field test of Ken's longtime dream of a dolphin sabbatical.

Ken had worked out the details of the sabbatical notion with Dr. Randy Wells, undoubtedly the leading authority on the bottlenose dolphins of Sarasota Bay. Randy had been studying the resident community of roughly one hundred dolphins for nearly twenty of his thirty-five years, starting as a high school student. Working with a growing cadre of scientific colleagues, plus a steady stream of Earthwatch volunteers (a program whereby people pay to spend their vacations participating in various research projects around the world), he has amassed an unparalleled wealth of information about this wild dolphin society. He knows who hangs out with whom, what parts of the bay they frequent, how far they travel, how fast they grow, who gave birth this year, who died or disappeared. He has followed three generations or more, watching youngsters grow up and have calves—even grandcalves—of their own. Out in a boat on the bay, Randy can often recognize individuals by the shape of their dorsal fins the way most of us recognize people by their faces.

We chose Tampa Bay as a capture site for the sabbatical project because of its proximity to Sarasota Bay. In fact, the two are interconnected: the northernmost tip of Sarasota Bay opens into the southern base of Tampa Bay. We didn't want to take dolphins out of Randy's study group but did want a location close enough so that his findings would apply. The discovery of pair bonds among Sarasota males, for example, led to our decision to go for two young males.

Randy was largely responsible for the day-to-day planning and organizing of the capture. He also acted as coordinator of the dolphin facility at Long Marine Laboratory, which would serve as the newly caught dolphins' home during their two- or three-year research stint. The lab was affiliated with the University of California at Santa Cruz (UCSC), where Ken taught and I was a graduate student in biology.

I'd flown to Florida three weeks earlier. We had transported Josephine and Arrow, the two dolphins we'd been working with previously, from Long Lab to a dolphin facility in the Florida Keys. Jo and Arrow were both older females, originally donated to us by the Navy—Navy surplus, I guess. Delightful characters both, but not good research subjects, as it turned out. Arrow had come to us with a variety of physical ailments and was never fully healthy; sweet of temperament, she nonetheless didn't seem to catch on to what we were asking of her. Josephine—well, Jo had a mind of her own, and doing research was not generally what she had in mind. After their lengthy tours of duty with the Navy—more than twenty years in Jo's case—the two of them were due for retirement, we figured. Where better to retire than Florida? We took them to the

Dolphin Research Center (DRC) on Grassy Key, where they would live in bay pens with other dolphins of their kind, be well fed, but not be required to work for their living.

We had considered releasing Josephine instead of retiring her, testing the "dolphin sabbatical" first with her. We didn't want to release her alone, however, and we had thus far been unable to find a suitable companion. Arrow simply wasn't healthy enough to be a promising candidate; she would need ongoing medical supervision and would be unlikely to survive in the wild. For a while we hoped an older male dolphin Jo had known in the Navy might accompany her, but the Navy ultimately said no.

Lack of a companion wasn't our only reservation. We were also concerned about Jo's age and the length of her stay in captivity. Would she remember how to fend for herself? For that matter, would she want to? Furthermore, we didn't know just where she'd originally been captured, so we couldn't return her to her home waters—an important aspect of the sabbatical notion. She wouldn't know the territory, nor the wild dolphin society in which she would find herself, and that could put her at a serious disadvantage. Perhaps most worrisome of all was that, when faced with new or frightening circumstances, Jo tended to quit eating, at least temporarily. To set her free without a familiar companion, in unfamiliar surroundings, could potentially be a death warrant. We decided, for the time being, to continue looking for an appropriate partner for Jo and to see how she adjusted to the DRC.

Michelle Jeffries, the head dolphin trainer at Long Lab, and I spent two weeks at the DRC, working with the staff there and trying to ease Jo and Arrow's transition to their new life. From there we headed for Sarasota, where we met up with Randy Wells and his group, who were at the tail end of one of their capture-release sessions. Randy runs several two-week research sessions each year. Some are survey sessions, simply monitoring the dolphin's behavior—who's where doing what with whom. One or two sessions a year involve actually catching some of the dolphins in the study area

—maybe twenty or thirty over the two-week period—taking a variety of on-the-spot measurements, blood samples and the like, and then releasing the animals, generally within twenty minutes to an hour.

Michelle and I joined them for the last couple days of the session. For Michelle it was her first glimpse of her husband-to-be in his element. She and Randy had met and fallen in love while working together with Jo and Arrow back at Long Lab. But clearly this was where his heart was; the dolphins of Sarasota Bay were his life's work. The two of them seemed a good match, and not just size-wise—Randy a shade over five foot six, Michelle pushing five one. The trainer and the field biologist each knew and loved dolphins, but in different ways, in different worlds. They complemented each other temperamentally as well—Michelle volatile and expressive, as befitted her Greek-Italian heritage; Randy the serious (though not solemn) scientist, intensely focused, quietly rational.

For both Michelle and me, this was our first taste of the capture process, and our introduction to the procedures we'd be using to catch the two new dolphins. No *Star-Trek*–style transporters here with which to beam the animals aboard. It takes some twenty people, four or five boats, and a fifteen-hundred-foot seine net.

First the boats spread out, communicating with one another by radio, in search of suitable dolphin groups. Randy drives the RV *Fandango*, a tiny two-person outboard that serves as scout boat. He is largely responsible for selecting which dolphins to go after. For each session he has his "wish list" of dolphins he especially wants to catch, individuals they haven't caught for several years (plus a few they've never been able to catch). They don't want to set their net on too large a group—more than a half-dozen or so animals could be unwieldy. Nor will they go after any group that includes a mother with a calf less than about two years old. They also try to avoid catching the more geriatric members of the community—those in their late thirties, forties, or even fifties.

When someone spots a promising group, the boats converge

and follow the dolphins until they are over a sandy spot of appropriate depth. Part of what makes Sarasota Bay particularly suitable for such a project is that much of it is less than six feet deep, so the animals can be caught in water shallow enough for people to stand up comfortably.

Larry Fulford, a local commercial fisherman, directs the actual capture proceedings. His vessel, the designated catch boat, is a twenty-five-foot "kickerboat," with the outboard motor set in the center instead of at the stern, allowing for greater maneuverability in shallow water and making it possible to cast a net off the back without getting it tangled in the propellers. Larry steers and calls out his directions from atop a six-foot sighting tower directly above the engine.

Once the boats are in position, Larry gives a signal, and one of the catch boat crew tosses out the "let go," an orange buoy attached to the end of the net. As the boat speeds up, the drag created by the "let go" pulls the net out behind it. Larry drives around the group of dolphins, encircling them within the net. Meanwhile, Michael Scott, one of Randy's principal collaborators in the Sarasota research, zips about in a Flare, a fast sports car of a boat, circling the dolphins in the opposite direction. This maneuver generates an "acoustic barrier," a wall of sound and bubbles that dolphins apparently are reluctant to pass through—perhaps because it interferes with their echolocation—preventing the animals from escaping before Larry completes the net enclosure.

Additional crew members stand by in one or two slightly larger boats correcting overlays (where the lead line that weights down the bottom of the net gets hung up on the float line at the top). Once the net is in place, most of the crew jumps into the water—praying not to land on any stingrays on the way in—and shuffles over to the net. People deploy themselves every ten feet or so around the outside of the net to discourage the dolphins from rushing into the net and to free them should any hit it.

Several of the more experienced members of the team then step inside the net to begin catching the animals individually, one at a time. Often they use a smaller "crowder net" to corral some of the dolphins into a more manageable space. Michael Scott generally makes the actual grab; not only is he most efficient at it, his gentleness tends to calm the animals. He takes hold of the dolphin from the side, just in front of its dorsal fin, with one arm over its back so he can hold it by both pectoral fins (front flippers). Andy Read, another scientist on Randy's regular crew, quickly falls in behind him. Several people hoist the animal into a small inflatable boat, where Randy and a veterinarian take the various measurements and samples and check the dolphin's condition. They then free the dolphin and repeat the process with the next animal.

Marine mammal veterinarian Dr. Jay Sweeney was helping Randy this session, as he'd done many times before. Swarthy, with a dark beard, a wiry build, and a touch of the rogue in his eyes, Jay has a certain swashbuckler air to him. He discovered a stingray barb deeply embedded in the side of one dolphin's head, just below the left eye. This wasn't the first dolphin they had found evidencing a close encounter with a stingray. While feeding in very shallow waters, the dolphins may drag a fin along the bottom and inadvertently stir up a stingray. Sometimes the stirring up is more deliberate: dolphins have also been known to toy with the rays, taking them in their mouths and tossing them about like Frisbees. Some such encounters may prove fatal for the dolphin, however, if the barb works its way into the heart or brain or other vital organ. Jay removed the barb before releasing the dolphin.

Most of the people who would form the catch team for the sabbatical project had worked with Randy before, either as staff or as volunteers, on these capture-release ventures. Jay Sweeney, a consultant to many oceanariums around the world, had led a number of previous dolphin captures and would serve as the "collector of record" for this one, supervising the operation. Larry Fulford would captain the catch boat. Michael Scott and Andy Read also agreed to help, though somewhat reluctantly; they were willing to catch dolphins, but they both clearly preferred releasing the animals right away to even the relatively short-term captivity we had in mind. Several other of Randy's old friends and colleagues offered their services, as did a few experienced Earthwatchers, including a postman from New Jersey who's spent every vacation he's had since 1985 working on the project.

Three of our people from Long Marine Lab also had prior experience on Randy's project and had assisted this past session. Howard Rhinehart was the lab's veterinary technician. His rather unimposing appearance—medium height and build, with sandy-gray hair and glasses—and his even-tempered, soft-spoken demeanor were the perfect foil for an outlandish sense of humor. Kim Urian and Jill Madden were both members of the crew that had cared for Jo and Arrow and would care for the two new dolphins when they reached Santa Cruz. For the last few weeks they had worked together as co-coordinators of Randy's Earthwatch volunteers. They had become such a twosome that everyone seemed to lose track of who was Kim and who was Jill, often using the names interchangeably, or calling them both Jim and Kill. Not that they really look alike, though both might be described as blond, petite, and pretty. Each is striking in her own way—Kim with a delicate, almost ethereal beauty, and a deep spirituality to match; Jill with a spectacular mane of golden hair and an expression that somehow manages to be feisty, mischievous, and acutely vulnerable all at once.

After the research session ended, those of us who had gathered at Sarasota drove up to Ruskin, on the northeast side of Tampa Bay, where we set up our base camp at the Bahia Beach Resort Motel. There we met up with the last few members of the team, including *National Geographic* photographer Flip Nicklin, noted for his photos of dolphins and whales, and Casey Silvey, part of the Long

Lab dolphin crew. An ample, buxom woman, generously endowed in both body and spirit, Casey greeted each of us with one of her trademark all-encompassing bear hugs.

After we'd unloaded our gear in one of the rooms, Casey and I headed out toward the pier in front of the motel. Casey stopped in midstride, grinned, and pointed up to the street sign on the road that ran along the end of the dock: Destiny Drive. "A sign from God?" she mused.

The local harbormaster, with the wonderfully appropriate name of Captain Thomas Sehorne (pronounced "sea horn"), had agreed to let us build a temporary holding pen for the dolphins alongside the pier. His sixteen-year-old son, Lee, immediately became an eager and invaluable member of our crew. We fashioned a rather crude-looking structure, with ten-foot lengths of PVC pipe glued together into rectangles, which we then lashed to the wharf pilings. We covered this framework with a two-inch mesh fishing net, secured to the pipes with plastic electrical ties. The finished product measured twenty by thirty feet, with a height of ten feet. Water depth in the pen varied with the tides, ranging from a low of about five feet to nine feet at high tide.

The next day, Monday, July 11, 1988, we set out looking for dolphins. I had thought Sarasota Bay seemed a good place to be a dolphin. The shallow, irregularly shaped bay, with all its cozy little mangrove-lined inlets, simply looked inviting. Tampa Bay seemed, to my eyes, far more intimidating. Not only was the bay itself bigger, but it sported bigger ships, bigger waves, bigger cities, and bigger industries on its shores. Deep shipping channels coursed the middle of the bay, where huge freighters and barges rumbled by. The dolphins apparently aren't intimidated by it all, however: more than a thousand of them make their home in Tampa Bay.

We spent most of Monday morning cruising along the coastline, searching for suitable-looking groups of dolphins in suitably shallow water. We wanted to find a group of subadults. Bottlenose dolphins, at least in this area, tend to form their own brand of youth gangs—not exactly violent, though they can get pretty rowdy. Once the calves are old enough to leave their mothers, usually around three to five years of age, they tend to hang out with a bunch of other kids about their own age. These youngsters engage in a lot of rough-and-tumble play, including a great deal of sexual activity. They will stay in such groups until they reach sexual maturity—at about ten to thirteen years of age for the males, somewhat younger for the females.

Mostly the dolphins we saw were either in the wrong age category or in water too deep for us to catch them. At one point, as we approached two very large dolphins, Randy said, "There's Jimmy Durante and Norman, and they're a long way from home." Jimmy Durante—so named because his misshapen dorsal fin resembles the late performer's great schnozz—and Norman formed one of the longest-standing male pairs from Sarasota Bay. Randy's crew had sighted them together consistently since 1980; wherever you saw one, you saw the other. This time they were well into Tampa Bay, about fifteen miles from their usual Sarasota haunts. That's one of the advantages of being part of a pair bond: such a duo can range much more widely than any single dolphin would dare.

We finally found a group that looked promising and cast the net around them. The crew jumped into the water and headed toward the net, to the strains of Captain Larry's "Shuffle your feet." As we reached the net, we discovered that only two dolphins remained; the others in the group had apparently slipped out before the net closed. We caught each of them individually, checked their genders, and measured them. The pair turned out to be a mother and a female calf, three years of age or so, judging by her size. Commercial aquariums and marine parks sometimes take older calves such as this one, figuring that the dependence the youngster still has on its mother may transfer to the trainer and thereby speed up the bonding process. Jay turned to Michelle, indicated the calf, and asked, "So what do you think?"

Michelle looked at Jay as though he'd offered her the business end of a stingray. "I'm not taking a baby from its mother," she told him. "Besides, we wanted two males—let's stick to the original plan."

Randy agreed. Since we were planning to release the animals again later, we wanted to be sure they already knew how to fend for themselves; we didn't want one that was still dependent on mom. We opened the net and watched the mother and daughter swim away.

We'd managed to avoid stingrays on that set but encountered instead another brand of "stingy thingies" as we came to call them: swarms of stinging jellyfish. The water around us teemed with the gelatinous, tentacled creatures. As we stood holding the two dolphins, the jellyfish stung us repeatedly, setting our legs, arms, backs, and bellies on fire.

We climbed back into the boats with no dolphins to show for our efforts but with plenty of stinging red welts rising all over our bodies. Conversation on the boat was devoted mostly to discussions of how to treat the stings. "Meat tenderizer," Howard suggested. "We can just sprinkle ourselves with Adolph's." "What, to make ourselves more palatable to the sharks?" I asked. No, Howard explained: jellyfish toxin is protein based, so anything that breaks down proteins should help. And that's precisely what meat tenderizer is designed to do. Unfortunately, nobody had thought to bring any along. Heat has a similar effect, so putting the affected areas in the hottest water you can stand should also take the sting out. Tampa Bay in mid-July may reach ninety degrees or so, but that's not hot enough to do the trick. The most commonly proffered remedy called for rubbing the stings with urine—also good at breaking down proteins apparently, and readily accessible. A somewhat embarrassing procedure to carry out under the circumstances, however. We just left our welts untended (and untenderized).

We continued farther down the coast and caught what turned out to be a single male off Port Manatee. He was larger than we were looking for, however, so we released him. By early afternoon a storm came up and blew us off the water. Because of Jay's schedule and our budget limitations, we had only a three-day time window in which to catch our two dolphins. One day down, and so far we remained empty-handed.

We set out again early the next morning. But the engine on Larry's boat kept overheating. By midmorning we had to head back to home base, empty-handed yet again. Efforts to repair the ailing motor failed. So his father's kickerboat, the *Little Terry* (generally referred to as the *Little Terror*, because it was a challenge to drive), which had functioned as our equipment and transport boat, now became the catch boat. We headed out again after lunch, with one less boat and several fewer people: those who weren't actively part of the catch teams had to stay ashore, including Casey, who was there primarily to take pictures, and a Florida Department of Natural Resources employee, who had been invited along as a courtesy.

This time we decided to head up into Old Tampa Bay, across the main bay from Ruskin, which meant crossing the shipping channel. At one point the boat I was in cut far too close, at least for my comfort, to a passing freighter. The wind was picking up, and with it the wave size, making for a jolting ride in our small craft.

I'd started my period that day, a definite inconvenience on this kind of venture. The men had a much easier time of it in general on these all-day journeys in small boats with no toilets. They could simply pee off the side if need be, a more difficult feat for the women in the crowd. The day before, Casey had reached the emergency stage and ended up hanging her backside off the stern while the boat slowed down a bit and the rest of us looked away. Sometimes, if time and place permitted, we could stop for a "swim break," where anyone who needed to could jump into the water and relieve themselves. Even that's of no use when you need to change a tampon, though. And pads weren't much good, given that we had to be ready to jump into the water at catch time. It was either stay back on the dock or risk making a bloody mess of things. I took the risk. I did wonder, though, if sharks are drawn to menstrual blood.

We finally encountered a group of dolphins along the Interbay Peninsula, offshore from MacDill Air Force Base. Signs posted in the water warn boats away from the area at the end of their runway. Dolphins pay little heed to such signs, however, and we followed where they led.

The group looked good—subadults, maybe half a dozen or so. Someone cast the "let go." Judging the number of dolphins in a group can be difficult, however, since at any given moment some of the animals may be under water. As we set the net around them, more and more dolphins began to surface. We'd caught not six but fourteen animals, more than we could handle with the number of people we had to guard the net. To make matters worse, we found we had at least one mother with a very young calf in the group. To top it off, the water was too deep—more like six to eight feet instead of the three or four we wanted. It's hard enough to hold a dolphin when you're standing; when you're in over your head, forget it.

We abandoned the set. People in the catch boat began hauling in the net as fast as they could. Others jumped into the water to keep the dolphins from getting tangled in the net as it was pulled in. The dolphins were rushing about, some of them hitting the net. Jill and Andy found the little calf with its tail flukes caught in the net. As they treaded water trying to untangle the baby, Jill felt something swoosh under her feet. Her eyes opened wide and she stared at Andy in horror. Sharks were Jill's greatest fear, and she was convinced that was what had just passed beneath their feet. "Don't think about it," Andy told her, "don't even think about it."

The group of dolphins swam out of the open net. One adult animal hovered nearby, however, as the calf was untangled from the net—presumably the mother waiting for her baby. The two of them swam off together as soon as the calf was freed.

Several of the other younger animals—past calf age but not yet fully grown—also stuck around, watching the operation. Curious youngsters, probably. Just the sort we were looking for. Once we got the net and all our people back into the boats, we set off again after this subset of animals. We tried to cast the net around the group, but they all slipped out before the net closed.

We continued following the subadult group as they led us through a tortuous channel system and around a sandbar into the mouth of Old Tampa Bay. The group even contained one animal clearly a bit smaller than the others—sure to be within our size range.

In the shallows just beyond a public swimming beach on Picnic Island, we set on the group a third time. This time we got them. The crew jumped into the water and headed for the net. No amount of foot shuffling would help with what we found when we got there, however. One whole segment of the net was simply plastered with stingrays, two or three dozen of them, all riled up from having a net land on top of them.

"Get them critters offa the net," Larry shouted down, singling out one of Randy's longtime volunteers on the capture-release program, whose muscular build had earned him the affectionate nickname Bubba. But brawn is no defense against an angry stingray. "I'm not going near there," Bubba responded.

First we managed to pinch off the net into a figure eight—part of the usual procedure, anyway, to confine the dolphins in a more manageable space—with the dolphins in one end of the net and the rays on the other. Larry himself then cleared the net, giving it a few good shakes and sending the battery of rays soaring off in search of a quieter stretch of sand. Fortunately, none of them soared into any of us.

We also discovered that instead of the four animals we thought we had in the net, we had eleven—still more dolphins than we'd bargained for, more than Randy and his group had ever caught at one time in Sarasota. To complicate matters further, the dolphins had been disturbed by the procedure of closing down the net and were roaring around en masse at high speed, sending up a wall of water and making it impossible to catch any one individual. At some point in the fray the little dolphin we'd been eyeing managed to jump clear of one of its companions and several people and sailed out over the net.

"There goes your animal," Jay observed as we all watched the little dolphin swim to its freedom. We still had ten animals in the net, however; now we had to catch each dolphin individually, check its gender, and measure its length to see if any of them was what we were looking for.

Another dolphin in the net, perhaps a bit smaller than the others, particularly caught Jill's eye—in fact, looked her right in the eye and held her gaze. "That one," she called out, pointing to "a goofy-looking dolphin with a tweaked dorsal fin," as she later described him. "That's a good one."

That one proved tough to catch, however.

I was one of the net guards, one of the dozen or so people stationed along the outside of the net, splashing the water in front of the net, if any dolphins started swimming our way, to keep them from charging into the net. We also had to be prepared to disentangle any dolphin that might hit the net.

While we were measuring the dolphins in the group, a marine patrol arrived to check on what we were up to. Apparently, some people at the swimming beach down the way had called the authorities when they saw us catching dolphins. We were pleased that people wouldn't just stand by and watch someone catch wild dolphins. While the rest of us were busy with the animals encircled in the net, Larry explained to the marine patrol who we were, where we were from, and what we were doing; he also gave him our permit numbers. Not just anybody can go out and catch himself a dolphin. Dolphins are covered by the Marine Mammal Protection Act, and the procedure for obtaining a permit to catch one is lengthy and complex.

Meanwhile Michael, Andy, Jay, and Randy stood within the net's compass, watching for an opportune moment to grab hold of one of the dolphins as they circled the enclosure. One by one they caught and measured the animals. A female. Not what we were after. She was slipped over the top of the net to her freedom. A male, but slightly over our upper size limit of 235 centimeters (about seven feet, eight inches) in length, measured from the tip of the snout to the trailing edge of the flukes. Again, over the net. One after another, the dolphins proved to be just a little too big, or the wrong gender. Finally only three animals were left in the net, including "that one," the one Jill had pointed out.

Randy checked "that one." A male. Two hundred thirty centimeters long. Perfect. We had a keeper at last. Our venture was no longer theoretical; we had a real, live dolphin on our hands. I started to cry. It's one thing to set off in search of wild dolphins; it's quite another to look one in the eye and say *you*, you there with the distinctive backgammon-board pattern on your forehead, you with the tip of your dorsal fin missing, your life has just changed.

I wasn't the only one in tears. We had all chosen to be there, but that didn't make it any easier. Those of us who would be working with the two dolphins in Santa Cruz had deliberately elected to take part in the capture ourselves, rather than simply pay to have it done, because we thought it important for us to feel the reality of what we were doing. Now it seemed all too real. It had been different somehow with Jo and Arrow. We had essentially inherited them from the Navy. They had both been in captivity many years before they came to us, and we had played no role in that. In fact, we liked to think we were able to offer them something of a respite from the rigors of Navy life.

This time, though, we were responsible. This time we were bringing the dolphins into captivity ourselves; we were consciously, willfully altering the course of their lives. This felt uncomfortably like playing God.

I thought about my role, about how much of the responsibility was mine. I was the one getting a Ph.D. out of the deal. Ken was the principal investigator, but it would be my dissertation. Sure, if I hadn't taken it on, Ken no doubt could have found any number of others more than willing to be his graduate student and carry out the project. An old standby of an excuse, but an impotent one. I had taken this on; I had chosen to take part. I had even helped make the decision to catch two wild dolphins.

When it became apparent that Jo and Arrow were simply not going to work as research subjects, we realized we'd have to turn elsewhere. We discussed the possibility of purchasing a couple of captive-born dolphins. (A moot point, really. Captive-born animals were too expensive for our budget.) Given that we wanted to study how dolphins use their echolocation, how they move and what they do as they echolocate, I told Ken, we needed dolphins who had made their living by echolocating, whose livelihoods depended on it. Dolphins born in captivity may still echolocate, but they've never had to do so to find supper, or to avoid being supper to a shark. Ken agreed.

The other two dolphins still in the net proved to be slightly over our size limit. We prepared to take the one "keeper" home. We placed him in a special dolphin stretcher, with holes for each of his pectoral fins and another through which he could urinate and defecate, if need be. Several of us carried him into shallower water and held him, in the stretcher, while others loaded the net back onto the boat. The dolphin whistled. He whistled again. No one answered. All his buddies had swum away. He was all alone in the hands of aliens.

I'm not sure just what the dolphins felt, but I know we were devastated. One of the volunteers, herself a former dolphin trainer, managed to lighten the mood somewhat by announcing "Our first doo-doo," cooing like a proud new mama at the dolphin's accomplishment. The rest of us laughed, the "coos" changing to "eews" as the brown, viscous goop swirled around our legs.

Twenty minutes later, the net on board, we were set to go. We started carrying the dolphin out to the catch boat, but as soon as he felt himself in deeper water, he tried to swim out of the sling. Several more people joined us to help carry and lift him onto the boat. We placed him on foam padding on the deck and spread A and D ointment (generally used to prevent or soothe diaper rash) over the dolphin's back and head to keep his skin from drying out. Now he even smelled like a baby's bottom. As further protection against drying out, we put a sheet over his back, with a slit cut for his dorsal fin, and then dampened the sheet. Seven people were stationed around the dolphin to keep him from thrashing and to keep him wet throughout the boat ride back. We had buckets of water and sponges, using the sponges to squeeze water gently over the dolphin's back, rather than just dowsing him with it. The dolphin continued to whistle occasionally.

It was after six P.M. before we set off toward Ruskin. The sky had clouded over and the wind had picked up. The waves had grown to a height of nearly three feet, making for a bouncy ride back. When we were about halfway across the bay, the storm overtook us. Driving rain. High winds. Lightning—and our boat was now the tallest thing in sight. Then the rain turned to hail. Dressed just in bathing suits and T-shirts, we shivered as the hail pelted down on us. The dolphin whistled. We leaned in over him, covering his body with our own to shield him from the icy pellets. As we finally approached the dock, the sky started to clear over Bahia Beach. Our return was heralded by not one but two rainbows in a spectacular double arch.

We pulled up to the dock next to the pen and waited a few minutes in the boat while the rain subsided. Jay took a blood sample from the dolphin. Then we lifted the dolphin, still in the sling, out of the boat and onto a foam pad on the dock. He remained quiet and still on the dock, with several people holding him, while the rest of us got into the water and positioned ourselves around the outside of the pen to make sure the dolphin wouldn't try to go through the net. It was low tide, with a water depth just over five feet in the pen; some of us could touch bottom, while others had to tread water or dangle from the side.

Jay and three others stationed themselves inside the pen to receive the dolphin as he was handed down by the people on the dock. Once the animal was in the water and freed from the stretcher, Jay held him in his arms and walked around the pen with him. He poked the dolphin's snout lightly up against the netting all around, and the people on the outside gently pushed the animal back away from the net, so he could learn the boundaries of his temporary home. After he'd been "shown the ropes," the dolphin was free to swim around and check out the area. He must have been totally baffled by the situation, or terrified; at first he wouldn't leave Jay's side, nestling up against him. Finally the dolphin dove under water. He surfaced face-to-face with Jill, and the two of them stared at each other once again. Jill pushed him back from the net and he dove again, disappearing into the murk. For the next minute or two we saw only the occasional air bubble rise to the surface. The dolphin came up for a quick breath and then went under again, seeming to stay down as long as possible. We dangled in the water outside the pen for maybe half an hour, ready to free him from the net if need be. He never hit the net again.

By the time Jay finally decided the dolphin was sufficiently settled and we could all get out of the water, the sun had just set, streaking the sky with crimson. Our moods were less sanguine. We seemed to be getting nearly every portent the heavens had to offer, a message as ambiguous and confusing as any Delphic oracle.

At last we were able to put on some dry clothes, warm up, and get some dinner. Eating was difficult, though. We took no joy in the day's catch. Most of us wondered if we could continue, if we could go out again the next day and catch another dolphin. Somehow, seeing that one lone dolphin in the pen was more than we could bear. Maybe if we had been able to catch two from the same group, two at the same time as we'd originally hoped, it might not have felt quite so bad; if there had been two dolphins in that pen, comforting and consoling each other, maybe it wouldn't have torn at our hearts—and our consciences—quite so strongly.

We wanted to have at least two people on the dock next to the dolphin pen at all times, including all night long, to make sure the dolphin was okay, and to keep people from bothering the animal. Michelle and Jill took the first night's watch, carrying their sleeping bags out and camping on the dock. They slept poorly. They couldn't see the dolphin in the darkness; they could only listen for his breath. Like a pair of young mothers, Michelle and Jill tuned in to the dolphin's breathing rhythms, and every time he deviated from his usual pattern, they awoke, holding their breath until the dolphin breathed again. The rest of us slept fitfully indoors.

I got up early the next morning and headed down to the dock. Yes, there was still a dolphin in the pen. I hadn't dreamed the whole thing. And we were about to go out to look for another one. Michelle and Jill stayed back with the dolphin we'd already caught as the rest of us set out in search of a companion for him. It was Wednesday, the third and final day of our catch window. This was it.

We headed south again, past Cockroach Bay and Port Manatee. By noon we still hadn't seen any suitable dolphins. Larry drove the *Little Terror* into Bishop's Harbor to check it out, while the other boats waited offshore. He came back out of the harbor accompanied by a group of eight dolphins. They appeared to be subadults. One of the dolphins leaped high into the air.

The depth was right, the bottom clear and sandy, the water calm. A perfect spot for a catch. We set the net. The catch team jumped into the water and shuffled toward the net. Jay announced that he'd seen a nine-foot bull shark in that very location just the day before. Great. Never mind the stingrays. Bull sharks are serious sharks—not like nurse sharks or some of the other wimpier varieties. Bull sharks have been known to attack humans. They are also one of the main predators on dolphins in the area. Randy has found that nearly a quarter of the Sarasota dolphins past calf age have shark-bite scars—and that represents only those who have survived such an attack—most of them probably inflicted by bull, tiger, and dusky sharks.

"Shuffle your feet, just shuffle your feet," Larry called out his refrain. "Stingrays and jellyfish and sharks, oh my," several of us chanted back, feeling like Dorothy on the road to Oz. "Stringrays and jellyfish and sharks, oh my." We definitely were *not* in Kansas.

We lined up around the outside of the net. Michael and Andy again went inside the net to catch each dolphin individually. Randy and Jay checked gender, condition, and length. We found we had two of the right size, one male and one female. Now we faced a dilemma. We hadn't made our ideal catch—two young males from the same group. We wanted two males, because that would be a more natural social unit for these dolphins, at least as they got older. We also wanted two we could be sure already knew each other, two who would get along together. We were out of time to keep looking for the ideal. So we could keep the male we already had back in the pen and this male in front of us, giving us the two-males part of the ideal. But they were caught in fairly distant parts of the bay from each other and wouldn't necessarily be acquainted. What if they hated each other? Forced to spend two or three years together in the same tank—we could be devising a dolphin version of Sartre's *No Exit*. We could, instead, take the two dolphins there in front of us. Males and females do hang out together as subadults, and since these two were traveling together, they clearly knew each other and presumably got along. They would be a less likely pairing as they approached adulthood, however, and so the two of them wouldn't necessarily form a natural social unit come release time.

Randy turned to me and asked, "What do you think?" I froze. Here we go again, I thought, playing God. Whose life shall we change? I was one of the people holding the female dolphin at the time. I was up front, by her head. I looked at her, stroked her. A beauty she was, her body taut, full of life, full of energy. She seemed to have a wildness to her I hadn't seen in the others. I was smitten. But what to do? I wanted very badly to take her home, to get to know her. I wanted to set her free, to let her live her own life. I'd make a most indecisive God.

I finally looked back at Randy and said, "Let's take her, let's take the pair." He agreed. So it was decided. We'd take these two back to the pen. We'd take the one we'd caught the day before back to the part of the bay where we'd found him and let him go. We loaded the two new dolphins onto the boat and headed back toward Bahia Beach. I stayed by the female. One of the volunteers, sitting across from me at the dolphin's head, was also smitten. He whistled softly to her the whole way back, making up pretty little tunes to calm her. She never did relax, though, thrashing her flukes several times during the trip, and keeping her eyes wide open the whole time. The male dolphin stayed quiet throughout.

We moored the boat with the two dolphins aboard to the dock next to the pen. Some of us stayed in the boat with the new dolphins, while others recaptured dolphin number one and took him out of the pen. He was set onto some foam padding on the dock, with several people holding him and keeping him wet. Then we prepared to introduce the two new animals, one at a time, to the pen.

The female went in first. Again, eight or ten of us deployed ourselves around the outside of the pen, while Jay and three others stayed inside it to receive the dolphin. The tide was much higher now, so none of us could stand this time. Jay swam the dolphin around inside the pen, holding her in his arms and showing her the netting, the boundaries of the pen.

That little female apparently wanted no part of all this. She started swinging her body back and forth to free herself, battering Jay with her beak and flukes in the process. Jay couldn't hold on. The dolphin must have detected—probably through echolocation—a hole in the netting, or at least a weak area in it. She headed for it full force, blasted through the net, and kept on going. The encounter left Jay with a broken tooth and a bruise the diameter of a basketball across his chest and shoulder.

We watched as the dolphin headed back out into the bay. And we wished her Godspeed. Bahia Beach isn't far from Bishop's Harbor, where we'd caught the pair, and it's within the range of the dolphins who frequent that area. That spunky little female should easily have found her way back to friends and family.

She had overridden our decision to go for two from the same group. I felt somehow I should have known better in the first place. We really had planned on two males, anyway. So that's what we now had. Maybe they didn't know each other, but then again, maybe they did—we didn't really know how widely these dolphins range, nor how much the different communities interact with each other. We were grateful, at least, that we hadn't already released the first male. We prayed the two would get along.

Randy checked the net where the female dolphin had gone through and found a small hole. It didn't look nearly big enough for a dolphin to pass through, though clearly it was. To cover that opening, and to try to eliminate the possibility of another such escape, we attached a second net, the one used as a crowder net in the catch proceedings, around the outside of the pen.

Meanwhile, the two remaining dolphins waited, the first-caught one on the dock, whistling and vocalizing loudly, the new one still on the boat, lying quietly. We first lowered dolphin number one, the "goofy-looking" one with the backgammon-board forehead, back into the water from his perch on the dock. Jay briefly reoriented him to the pen before releasing him. Then we brought in the other male from the boat, and Jay gave him a similar introduction to the place, swimming him around the pen and showing him its boundaries. When Jay released him, the second dolphin dove under water and immediately surfaced beside the first animal. The two of them swam around the pen together, side by side, pectoral fin to pectoral fin.

It was after three P.M. by the time we got both animals into the pen, and we suddenly realized we were starving—we'd had no time for the lunches we'd so carefully packed away. Most of us were still dangling in the water around the outside of the pen, trying to ensure that our two dolphins would stay put. Someone in the boat started handing out peanut-butter-and-jelly sandwiches, which we munched on as we treaded water. So much for waiting for an hour after lunch before going swimming.

The two dolphins seemed to settle down together. After about an hour Jay decided we could end our watery vigil. Casey, Jill, and I were about to swim in the thirty yards or so to shore when Michael called over to us, "Get in the boat." We all felt more like swimming at that point. Besides, I've never been particularly adept at hoisting myself onto things and was none too eager to display my ineptness. Casey was even less eager to clamber aboard the boat than I was. We ignored Michael and started toward shore. He repeated, "Get in the boat," this time with an urgency, a tenseness, in his voice that caught my attention. I convinced Casey, who was still determined to swim in, and Jill that we'd better do as he said. Casey and I were hauled rather gracelessly up onto the boat. Then Jill, who'd gallantly allowed us to go first, climbed aboard. She would've shown no such courtesy—in fact, would have run right up our backs, Jill said later—if she'd known what Michael had seen: a shark. A black-tip shark, to be precise, cruising the waters between us and the shore, apparently feeding on mullet. A fairly small one, Michael assured us—only five feet long or so. Still, best not to come between a shark and its lunch. I was starting to wish we *were* in Kansas.

Now we had our two dolphins. We still wanted to keep them in the temporary pens for a week or ten days to make sure they were adjusting to captive life before we set off for California with them.

The most crucial step in the process is getting the dolphins to accept the dead fish that will be their fare in captivity. In the wild, of course, they're accustomed to catching their own meals, alive and aswimming. In fact, a wild dolphin will generally avoid dead fish— who knows what the fish may have died of, or how long it's been that way? Some newly caught dolphins may refuse dead fish for weeks and have to be force-fed to keep them alive until they get the idea. A few never adjust and must be released again. We wanted to keep these two dolphins close to home until we were certain they would eat.

The usual procedure is to work gradually from live fish to less and less lively ones until, before they know it, the dolphins are eating dead fish. Unfortunately, that means disabling the fish, clipping off their fins so they will flop about in the water, attracting the dolphins' attention and serving as easy prey. I couldn't quite shake the image of little boys pulling wings off flies.

We wanted to try mullet, a common fish in the area and a staple of the local dolphins' diet, but couldn't find any place that sold them. So we started off with pinfish and river shad, purchased at a local bait shop. On Wednesday, the day after his capture, Michelle and Jill had offered our first-caught dolphin thirty such fish. He refused them all. We hoped that when we brought in the second dolphin later that day they would both start eating. Neither ate a single fish, disabled or otherwise.

On Thursday morning a local deputy sheriff stopped by to check our credentials and see what we were up to. He ended up offering his services. His name was Hutch, he said, and he just happened to own a cast net for catching mullet. He was well practiced with it, too. He came back that afternoon with a garbage can full of water and mullet. Michelle picked out a couple that were still alive, but just barely, and tossed them into the dolphins' pen. A swirl of dolphins and water, and the fish were gone. Michelle threw in two more mullet, this time dead ones. Those, too, disappeared. Michelle kept tossing fish, and the dolphins kept eating them. They downed fifteen mullet between them at that first supper. Hallelujah. We danced on the dock. There would be no force-feeding with these two. And no more need to maim defenseless fishes.

Hutch gave generously of his time and mullet over the next few days. He had other things to do besides be our resident fisherman, however, so he loaned us his net and tried to teach Michelle and me how to use it. It's a lot trickier than it looks. You have to drape part of the net over one arm somehow and toss it with the other arm— preferably without hurling yourself into the water after it. Neither Michelle nor I ever landed anything but a few chuckles from our peers. Hutch continued as our supplier.

Michelle, Howard, and Lee went into the pen periodically with scuba gear to pick up dead fish off the bottom so we could get rid of any rotten ones, and so we could try to keep track of how many fish the dolphins were eating. We all took turns getting into the pen and swimming with the dolphins. Actually, we didn't so much swim as float about and let the dolphins decide whether they wanted to come near or not. Mostly they didn't. At first they would swim to the corner and face away from the person in the pen, hoping the strange creature would go away, I imagine. Sometimes they seemed to get agitated, slapping their tails on the water. We'd take the hint and get out. Gradually they got bolder, though, and would make an occasional quick pass by one of us, checking us out with their echolocation.

The two dolphins did seem to hit it off with one another, or at least they turned to one another for some sort of comfort and familiarity. Even if they didn't know each other as individuals, at least they knew they were of the same kind—unlike these alien beings throwing dead mullet at them from the dock.

The two of them stayed close together, and especially at first they seemed to synchronize their movements. They'd swim together, surface together, take a breath together. Initially, they mostly stayed in the far corner of the pen, away from the dock, away from the people, except at feeding time. Gradually, over the course of a few days, they came closer to the dock. They became less synchronous in their movements. They started poking their heads up out of the water, looking at us.

We heard rumors that some people in the area didn't like what we were doing. There was even talk of someone sneaking out and cutting the nets, freeing the two dolphins. So after the second dolphin arrived, we set up camp on the dock in earnest, with a tarp over a part of it to protect us and our gear from the sun and from the frequent outbursts of rain. At least two of our crew members were on dock duty throughout the day. We increased our nighttime guard to fully half the crew, with everybody alternating nights. We would all haul our sleeping bags or foam mats and blankets onto the dock and sleep out under the stars, sort of a dockside slumber party—though we weren't exactly in a partying mood.

Wednesday night, the first night with both dolphins in the pen, was a new moon, so when the bar and restaurant closed down for the evening, the shore disappeared into the darkness. With only water around us and stars above, we seemed almost to be adrift at sea. From time to time the lapping of the waves was punctuated by whistles and chatter from the two dolphins. I wished I knew what they were saying to one another. At one point some wild dolphins swam by not far from the dock, causing great consternation in our penned-in pair. Their plaintive whistles haunted our dreams.

I woke very early the next morning, a little before sunrise. Flip was already up and taking pictures in the light of dawn—the dolphins swimming quietly in their pen, the motley array of bodies bundled in blankets and sleeping bags strewn about the dock, all covered with the early-morning dew.

Leaving two people on guard with the dolphins, the rest of us trooped over to the restaurant of the Bahia Beach Resort for breakfast. We put in our orders. We waited. Other diners got their meals and left, and still we waited. We complained. Finally our breakfasts came. The toast was badly burned. The eggs and the potatoes were positively swimming in grease. Everything was stone cold. All inedible. We heard later that the cook disapproved of our catching dolphins; apparently she'd used the materials at hand to signal her displeasure. We got the message.

From then on we drove into the town of Ruskin, to a small café called the Coffee Cup—a local favorite, it seemed, where the townspeople gathered to eat and chat. It was the kind of place where you could get lunch specials of fried chicken or meat loaf and potatoes and vegetables and God knows what, all for $3.25. Breakfast cost $1.10 for two eggs, toast, and grits (you could get hash browns instead, on request). Grits were a new experience for many of us, especially the California contingent. We were accustomed to seeing alfalfa sprouts on our plates regardless of what we ordered; omnipresent grits were another matter. But the food was good as well as cheap, and the people were friendly. No one there ever asked why we were stealing "their" dolphins.

Some of the people who came out onto the dock to see the dolphins did ask precisely that, and other questions as well. Local residents, visitors at the resort, people who ate at the restaurant or frequented the bar there, would often wander out to see what was going on. Most were friendly or just curious; a few were fascinated and visited regularly. We did our best to answer all their questions, amicable or otherwise.

"Don't you have dolphins in California?" several people asked. Yes, we do, several species. There's even a variety of bottlenose dolphin—the kind we were after. Part of the reason we were in Florida was simply that the shallow bays make it easier and safer to catch the dolphins there than along the coast of California. In fact, most of the wild-caught bottlenose dolphins in oceanariums, aquariums, and research facilities in this country have probably been caught along the Gulf coasts of Florida, Texas, and Mississippi.

More important, though, we chose Tampa Bay, Florida, because so much was already known about the dolphins in the adjacent Sarasota Bay. Furthermore, Randy and his colleagues had recently been awarded a contract from the National Marine Fisheries Service to carry out a photographic identification census of the dolphins in Tampa Bay. That project would coincide nicely with the release and follow-up study we had planned with these two newly caught dolphins; it meant we'd know more about the dolphin society from which these two came and to which they would return.

My favorite dockside visitor was the girl with the sunburst eyes. About eight years old, with curly blond hair down to the middle of her back, she had eyes the likes of which I've never seen before or since—deep blue with shafts of gold radiating outward from the pupil, like the noontime sun in a cloudless sky. She came daily to visit the dolphins, sometimes asking questions, but mostly just sitting silently on the dock, those wondrous eyes shining down on the two dolphins in the pen below.

Other visitors were rather less agreeable. Late Friday night Randy stood guard while the rest of us dozed on the dock. He woke us with a loud whisper: "Somebody's coming." We heard some splashing in the water, a muffled voice. This is it, we thought. They've come to cut the net. What would we do against a band of would-be saboteurs, anyway? We were just scientists and students, not soldiers. I felt as though I'd fallen asleep during *Flipper* and woken up to a John Wayne movie. But as we listened more closely to the erratic splashing and the unintelligible prattle, we realized this was no sabotage team—just some lone drunk who'd decided on a late-night swim after the bar closed. "Lucky he didn't end up shark bait," Randy muttered as we settled uneasily back into our sleeping bags.

Another after-hours caller stumbled out onto the dock two nights later. This one was even drunker and more loquacious than the first. He turned out to be harmless enough, though, and most of us fell back to sleep. Casey, kind-hearted soul, sat up with him most of the night, patiently listening to repeated tellings of his story. He was there, he explained, commemorating the two-year anniversary of his brother's death. Two years ago to the day, at that very beach, his brother had been killed by a blacktip shark—the kind we'd been ordered onto the boat to avoid a few days earlier.

Late Saturday afternoon, just about suppertime, one of the crew members staying in the room next door came over and told me I had a phone call. The caller had asked first for Randy or Michelle or Howard, all of whom were out. I headed for the phone with some apprehension—what sort of news would I be fourth in line to hear?

I picked up the phone. It was Jane Rodriguez from the Dolphin Research Center, where we'd taken Jo and Arrow. Jane's voice only heightened my sense of foreboding, and her words confirmed it: "Arrow just died." "We did everything we could," she assured me several times. "I know you did," I reassured her as many times, talking through my tears. Jo had stayed right with Arrow at the end, she told me, stroking her and helping her stay afloat. Good old Josephine. A crusty old soul, but tenderhearted. They were about to perform a necropsy on Arrow (the procedure is called an autopsy if performed on a member of our own species, a necropsy if it's another species), Jane said, and would let us know the results.

As Randy and Michelle returned from dinner with his parents, and Howard came in from a long walk, they faced a gloomy, teary-eyed crew. I told them of Jane's call. I'd never before been the bearer of such news, never had to report the death of someone I loved, someone we all loved. Arrow's death hit us like a death in the family. Michelle and Howard, who together had accompanied Jo and Arrow on the flight to Florida, just stood holding each other, rocking slightly, too distraught even to cry.

The necropsy pointed to "nonspecific organ failure" as the cause of death. We knew Arrow had chronic liver problems and recurring urinary-tract infections, but just why she died at this point was unclear. We'd given Jo and Arrow complete physical exams just prior to the trip to Florida, and both were certified fit for transport—Arrow had been at her healthiest. We'd done everything we could to make the transport and the transition as smooth as possible for both dolphins. Arrow had seemed fine on the trip—in fact, she took the whole thing considerably more calmly than Josephine did. She'd even carried on an extended "conversation" with some chickens who were on the same transport flight, whistling and clucking back and forth with them. We discovered when taking her out of the transporter that she'd developed some blisters along the crook of her front flippers (the "armpit" area), not uncommon in transports. Jo had similar but smaller sores. They otherwise seemed to have come through in good shape. Still, moving is stressful; perhaps the cross-country journey and the new environment were simply too much for Arrow's rather delicate constitution. She hadn't been eating well at the DRC, but then neither had Jo. The Florida water was nearly thirty degrees warmer than the tank back in Santa Cruz, however, so we figured they may simply have been trying to lose their cold-water blubber layer. Both dolphins had seemed to be on the upswing when Michelle and I left the DRC for Sarasota.

Arrow's death cast an even greater pall over our current endeavor. We couldn't be sure to what extent, if any, it might have been captivity related. We wondered all the more just what we were bringing these two new dolphins into. And we wished all the more that Ken Norris were there, to help strengthen our resolve, I guess—to remind us why we were doing this, to offer his own special brand of Grand Pooh-bahing. He was stuck back in Santa Cruz dealing with some bureaucratic matters. Most of us were at least a little angry at Ken, nonetheless, for not being with us for the capture. This was, after all, his project; he was captain of this ship, yet he wasn't even aboard. With Arrow's death we felt we were sinking.

Randy called Ken that night to let him know what had happened; he told Ken, too, of the crew's disaffection. Ken called the next morning to talk to us, to try to offer some solace long-distance. I don't remember feeling much comforted.

I later tried to explain to Ken what we'd been feeling. He truly didn't understand why we were all so upset by catching dolphins in the first place, perhaps because he was so aware of how much better conditions are now than when he'd started out. Attitudes toward dolphins have changed so dramatically in the nearly forty years since he'd got into the business—a time when people would harpoon the animals for sport, when many researchers would readily "sacrifice" them in the name of science. Ken himself played a major role in effecting those changes, helping to bring about the Marine Mammal Protection Act of 1972, to institute bans on whaling, and to establish guidelines for the tuna industry to reduce the number of dolphins killed each year in their fishing nets.

We were a rather dispirited crew that Sunday, the day before we were to send the dolphins on their way to California. We'd decided to shorten the dolphins' adjustment period from the ten days we'd originally planned to just under a week, in part because the dolphins were doing so well already, and in part to avoid any further threats or harassment.

Things were quiet on the dock Monday afternoon. Kim Urian and I were standing watch. Everyone else was back in the rooms or out making final preparations for transport. The sky started to cloud over and the wind picked up suddenly. Clearly we weren't going to make it out of there without yet another storm—and a fearsome one at that. Gale-force winds battered us and ripped the tarp that provided our only protection from the elements. I stood up, and the lawn chair I'd been sitting on sailed off across the bay. I very nearly followed it. Kansas was looking better by the minute.

Kim was wearing her camera around her neck, and it apparently went off accidentally at some point. She ended up with a picture of me, taken at a skewed angle, with my eyes wide and wild, my hair standing on end from some combination of wind, storm-generated electricity in the air, and plain old fright.

The wind and waves battered the dolphin pen and began to tear it apart in one corner. Kim ran back to the rooms to get help. She glanced up at a nearby flagpole as she ran by; the flag had been shredded by the wind. She found herself praying, pleading with God, promising never to complain again if only we all might survive the afternoon.

I stayed on the dock near the dolphins, lying low to avoid being blown away, and, I hoped, to avoid being struck by the lightning that was flashing all around us. Kim returned with Randy, Flip, Lee, and Michelle. Flip and Lee jumped into the water and hastily lashed the pen framework back together.

The others went back to the rooms to wait out the storm. Randy, Michelle, and I took shelter in a small gazebo on shore near the end of the dock. We trusted that the flagpole, not the gazebo, would attract any lightning strikes in the vicinity. Our role, if any at that point, wouldn't be to try to catch or keep the two dolphins, but to free them if need be. If the storm breached the pen, we had no hope of holding on to the dolphins; but we would need to make sure the netting didn't collapse in on them and trap them underwater—dolphins die without air just as surely as humans do. I doubted that any of our crew had the heart or the will to catch more dolphins, should these two go free now. The whole project seemed to hang in the balance, in the fury of that storm.

Randy, Michelle, and I, the three whose lives would perhaps be most intimately intertwined with those of the two dolphins in that pen, huddled together in the little gazebo while the storm raged around us. Soaking wet and shivering, arms around one another, we watched as the sea pen rattled and shook, wondering if it would hold, wondering if we wanted it to.

2

Life with Humans

*T*he pen held.

The storm subsided slightly, and Randy, Michelle, and I finally ventured out from the shelter of the gazebo to begin preparing for the transport back to California. It was after four P.M. already, and the dolphins had an eleven o'clock plane to catch.

Kim and Michelle started gathering supplies for the transport. We would need an assortment of sponges and sprayers, several large jugs of water, and plenty of ice to keep the animals cool and wet during the trip—one of the major concerns in transporting dolphins. The ice maker was outside, unsheltered, and lightning still flashed too often and too close for comfort. Kim and Michelle touched the metal machine gingerly, trying to scoop ice between lightning bolts, certain they'd be frizzled at any moment. Kim made more promises.

Others of us finished preparing the transporter devices the dolphins would ride in. Each transporter consisted of a rubberized canvas bag fitted onto a metal frame. We wrapped all metal parts of the transporters in foam to prevent the animals from hurting themselves if they should start to thrash during the trip. The dolphins would ride in a stretcher or sling, with the sling poles resting on top of the metal transporter frame. Ice water would go between the sling and the canvas bag.

Proper weight support is another major concern in transports. Dolphins are built for the buoyancy of water; out of water, their own body weight begins to press down on them, putting pressure on their vital organs. The sling would do part of the job. The bulk of the weight support, however, would be provided by four-inch-thick sheets of foam that lined the canvas bags, with holes cut in the foam to accommodate the animals' pectoral fins.

By six P.M. we were ready to begin moving the dolphins. The wind and waves had abated, the rain dwindled to a drizzle. As we walked out onto the dock, the two dolphins looked up at us and started chuffing—a huffy, indignant sort of sound produced by expelling air forcefully through the blowhole, apparently used to indicate a degree of anger or frustration. To ease the transport, we hadn't fed them that day: we didn't want them lying on full stomachs the whole way, nor did we want them lying in their own excrement. They were undoubtedly hungry and probably mystified. Here we'd just convinced them to eat dead fish, and then we stopped feeding them. Capricious creatures, these humans.

We decided to take the second-caught dolphin from the pen first, because he seemed the calmer of the two. Four people got into the water and caught the animal, using a crowder net to force him to one corner. They placed him in the sling and handed him up to those of us on the dock. We carried the dolphin to a fifteen-foot enclosed Ryder truck parked near the end of the dock. Once he was placed in the transporter, we again applied A and D ointment to all exposed areas and draped a wet sheet over his back to keep his skin from drying out. Michelle and I stayed with him while the rest of the crew went back for the other animal. We sprayed him with water and kept track of his breathing rate. We would also try to mollify him should he start to struggle. He didn't. He remained quiet and calm, breathing regularly, never making a sound. He just watched, following our every move, with eyes that seemed unusually large for a bottlenose dolphin.

The crew brought in the other dolphin and placed him in an identical transporter, next to the first. He, too, got the A and D treatment and a sheet over his back. This one was very vocal, whistling and making noises, struggling occasionally. He kept his eyes tightly shut. Jill stayed by him—the one she'd taken to from the outset—and kept up a whistling exchange, the two of them whistling back and forth to each other throughout the truck ride.

We drove two hours, traversing most of the width of Florida, six of us in the back of the truck with the dolphins, the others following in cars. We'd booked passage out of Orlando through Flying Tigers, an air-freight company. When we arrived at the terminal, the dolphins were removed from the truck by forklift, with people arrayed all around each transporter to support it while it was being lowered. Before loading them onto the pallet, we removed the dolphins from their transporters and repositioned them, making sure no part of them was rubbing against the transporters. We also cut away some of the foam around the dolphins' heads to keep it from blocking their vision; they should be able to see what was going on, if they wanted. The quiet one with the big eyes wanted; the whistler did not.

Jay Sweeney and Howard Rhinehart, our vet tech, would accompany the dolphins on the flight to California. The original plan had called for Howard and Michelle, but following Arrow's death, we wanted to take every possible precaution, including having a full-fledged veterinarian on the transport. Randy asked Jay, who had already returned home to San Diego, to come back to Florida. He agreed to fly across the country once more—and then turn around and fly back again. Since this was a cargo flight, it couldn't accommodate more than two people. In fact, the two dolphins in their transporters, all the necessary equipment—water and ice, sprayers, sponges, extra foam, medical kits, Howard's and Jay's personal luggage—and the two people had to be packed onto a ten-foot-by-ten-foot pallet. Narrow aisles between and around the two transporters would be all the space Jay and Howard would have to maneuver in; they would be so much cargo themselves.

We placed the two transporters onto the pallet, and the freight handlers lashed them into place. Then we waited. The Flying Tigers crew loaded everything else onto the plane first; the two dolphins would be the last aboard. We spoke softly to the dolphins and sprayed them with water regularly, keeping them wet and as comfortable as possible. Both seemed amazingly calm, or perhaps simply too overwhelmed to respond. Finally it was their turn. Our crew watched and waved solemnly as the freight crew pushed the pallet with the two dolphins down a long set of rollers and out the cargo door. Another forklift hoisted them onto the plane. It was eleven P.M.

The flight followed Flying Tigers' normal cargo route, complete with a two-hour layover in Cleveland. The cargo hold was pressurized but not heated, which was better for the dolphins—they'd be less likely to overheat—but "butt cold" for the humans, Howard said. I doubt the flight was very comfortable for any of the four animate cargo occupants. Howard and Jay strapped themselves into a pair of jump seats for takeoffs and landings but otherwise spent their time standing over the dolphins, spraying them, adjusting them, watching them. The two dolphins remained tranquil almost till the end. Shortly before they arrived in San Francisco, the quiet one—the one who had seemed the calmer of the two—began to breathe rapidly, in distressed fashion. Jay injected him with dexamethasone, a steroid, which calmed the dolphin and corrected his breathing difficulties. The plane touched down at the San Francisco airport about seven A.M.

Holly Green, one of our regular dolphin crew members who hadn't taken part in the capture, had been responsible for preparing the tanks at Long Marine Lab for the dolphins' arrival and for organizing a receiving team at the airport. She and half a dozen others from the lab met the plane at the Flying Tigers terminal. They loaded the travel-weary dolphins onto yet another Ryder truck for the final leg of their journey, yet another two-hour ride, this time through the Santa Cruz mountains to the Pacific coast. Shortly after ten A.M. on Tuesday, July 19, the truck pulled up through the Brussels sprout fields that surround Long Lab, through the dense fog that frequently enshrouds it, and delivered the two dolphins to their new home.

Long Marine Lab is set amid coastal farmlands on the outskirts of Santa Cruz, perched atop a cliff overlooking the Pacific Ocean at the northernmost lip of Monterey Bay. Officially, it's the Joseph M. Long Marine Laboratory, named for the founder of the Long's drugstore chain, who donated most of the funds to build the facility. Informally, it's known as LML. It consistently boasts the coldest, windiest, foggiest weather in town. Even when the rest of Santa Cruz is basking in ninety-degree sunshine, the fog bank seems to hover right over the lab. You can see it there waiting for you as soon as you reach the entrance gate; the temperature drops twenty-five degrees as you drive in. Novice visitors to the lab, thinking the weather should remain consistent from one end of town to the other, often arrive in shorts and tank tops and freeze their tushies off. Seasoned veterans bring layers of extra clothing.

LML's dolphin facility was just two years old, having been completed in June of 1986; Jo and Arrow had been its first and only occupants thus far. The outdoor three-tank complex consisted of a large ovate pool with two smaller circular pools attached to one end. If viewed from above, it looked a bit like Mickey Mouse—an oval face with two round ears. Make that a fifty-by-forty-foot face with one thirty-foot ear and one twenty-five-foot ear.

The exact dimensions of the large pool are difficult to specify, because Ken and Randy had designed it with tank acoustics in mind. To reduce reverberation problems, they gave the tank an irregular shape, with sloping walls and no two surfaces exactly parallel to each other, so that sounds made within wouldn't just bounce straight back at whoever was making them. That way the dolphins wouldn't feel as though they were in an echo chamber.

The fifty-foot tank held some ninety thousand gallons of water. It was connected to each of the smaller pools by a four-foot-wide opening, which we could close off either with a simple net gate or with a watertight gate to separate the tanks completely. The two smaller tanks were connected to each other by a long channel, which had been intended as an area to do medical work on the dolphins but which, in fact, saw little action of that sort. (For one thing, it was too narrow; one good flick of the dolphin's flukes, and anyone holding him would be plastered to the wall.) An underwater viewing room ran along one side of the tank complex below the surface, with four windows looking into the large tank and one into the thirty-foot pool.

Holly had filled the tanks and turned on the heater in anticipation of the new dolphins' arrival. Unlike inland facilities, which generally have to manufacture their salt water, LML's tanks held actual seawater, pumped up from the ocean at the bottom of the cliff and stored in two large silos next to the tanks. It was lightly chlorinated—at about one tenth the level people use in swimming pools—in a not altogether successful effort to reduce the algae growth. The seawater ran continuously through a filtering system and was recycled back into the tank, along with a small, steady inflow of water fresh from the silos.

The heater allowed us to warm the water somewhat above the fifty to fifty-five degrees Fahrenheit typical of Monterey Bay, though we'd never be able to match the temperatures the dolphins had just left. In Tampa Bay the water ranges from about fifty degrees in winter to ninety-plus in summer. The newly arrived dolphins were in for an early autumn—the tank temperature registered a balmy seventy.

The Ryder truck backed into the fenced compound surrounding the dolphin tanks. The crew, now joined by several more lab people, removed the transporters from the truck by forklift and carried the dolphins one by one in their stretchers to the side of the large pool. The one who'd had the breathing difficulties on the plane went first. The crew members set him down on the deck at the very edge of the pool, lowered the side of the stretcher next to the water, making sure his pectoral fins were free, and gently rolled the dolphin into the tank. He hit the water swimming. The other one—Jill's whistler—sank immediately to the bottom of the tank and lay there. No one breathed. Holly prepared to dive in after him, if need be. A moment later he surfaced, exhaled loudly, and joined his tankmate. The two dolphins began to swim together in tight formation. The humans resumed breathing.

Howard brought out buckets of herring—a type of fish these two dolphins probably never had seen before in live form, much less dead and previously frozen—and started to toss fish into the water. To everyone's amazement and delight, the pair ate hungrily, devouring all the fish they were offered.

Meanwhile, the rest of the crew made our way back from Ruskin. Kim, Jill, Randy, and I stayed in Florida for a few days after the dolphins left to take down the pen, pack up, and put things away. Casey and Michelle, who simply couldn't bear not being on the transport flight with the animals, caught the earliest possible flight back to San Francisco Tuesday morning. If Michelle could have made the pilot fly faster, she would have. She had no patience for the flight, nor for the two-hour drive from the airport to Santa Cruz. They ran into a traffic jam along Highway 17, a hazardous, winding mountain road that serves as the major artery connecting Santa Cruz to the outside world. Michelle cursed the traffic and prayed for the dolphins. She hadn't yet heard how the transport had gone, or whether the dolphins had arrived safely.

Michelle and Casey pulled into Santa Cruz about three P.M. and headed straight for LML. Never mind that Michelle hadn't been home in a month, or that she'd barely slept the night before; she wouldn't rest until she saw the dolphins. People at the lab were so calm, it was almost anticlimactic, Michelle said later. "The dolphins are doing fine," they told her. To satisfy herself on that score, however, Michelle immediately grabbed a bucket of fish and ran out to the tank. Both dolphins ate heartily, this time consuming capelin—yet another novelty for them—along with the herring. Michelle, always happiest when feeding animals of one sort or another, finally began to relax.

I wish I knew what the dolphins made of the whole experience. Nothing they'd learned from their mothers, or from their schoolmates back in Tampa Bay, could have begun to prepare them for such a discordancy of strange new sights and sounds and sensations. Not even tabloid accounts of abductions by extraterrestrials describe anything more alien than what these two had encountered. Fully aquatic animals, they were lifted completely out of the water, out of their element. Animals accustomed to being always on the move, they were immobilized, their bodies restrained, confined in a box.

They were transported about in so many mysterious ways—bounced along in a truck, lifted by forklifts, pushed along on rollers, swept off in an airplane. They were subjected to the feel of the stretcher, of A and D ointment, of human hands against their skin. The glaring, unnatural lights of the airport terminal. The incessant clatter and grind of machinery. The unintelligible chatter of all those humans. Everything alien. Aliens everywhere.

Would the dolphins even remember the experience later? Would they have the context, the framework, from which to form memories? Or would it all be lost to memory, as birth is for us? The transport ordeal in some ways did parallel the birth process: physical constriction, unfamiliar motions, emergence from fluid into air, the assault on the senses. In a way, they truly were born again, reborn as captive dolphins. From the moment they landed in the tank at Long Marine Lab and ate their first herring, they started a new life.

They got new names, too, to go with that new life. They continued to call to each other with their whistles. Each dolphin, at least of this species, apparently has its own unique "signature whistle," which functions something like a name. But we humans needed human-type names to call them. Though we'd already settled on names back in Florida, they really didn't seem official until the dolphins arrived in Santa Cruz. In Ruskin, even in the pen, the dolphins were wild animals still; the names had no hold on them. At Long Lab, as the dolphins settled into life with humans, the names accrued meaning.

Names for the two dolphins had been the subject of some debate. For the first-caught one, the one brought in amid storm and rainbows, I suggested "Prospero," after the magician-scholar in Shakespeare's *The Tempest*. Casey vetoed the idea. Prospero is also the name of a produce market in Santa Cruz, whose somewhat abstruse sign out front depicts a man in full academic garb, complete with mortarboard cap and flowing black gown, touching a magic wand to a giant mushroom. "Prospero makes me think of mushrooms," Casey protested, effectively ending the discussion.

We briefly considered calling him Gammon, because of the backgammon-board pattern on his forehead, but that too was rejected. I've since forgotten why and by whom. We agreed finally on a name we'd been considering even before capture: Echo, short for echolocation, which was to be the focus of our studies with the two dolphins.

I discovered later that the name has a rather troubling history: in Greek mythology Echo was a lovely young nymph whose lively chatter attracted the attention of Hera, Zeus's wife, on one of her frequent jealous rampages. The goddess condemned Echo never to be able to speak except to repeat what was said to her. "You will always have the last word," Hera proclaimed, "but no power to speak first." As if that weren't bad enough, Echo had the further misfortune to fall in love with Narcissus, the ineffably beautiful youth who had eyes only for himself; the nymph lived out her days not only echolalic but brokenhearted. Fortunately, the dolphin Echo proved himself fully capable of speaking first, as well as last.

In keeping with the echolocation theme, most of the crew wanted to call the second dolphin Click. Dolphin echolocation sounds are composed of a series of brief clicks; strung together in rapid sequence, these click trains sound something like a creaky door. By listening to the returning echoes of the clicks they produce, dolphins can form highly complex acoustic images of their surroundings. You can't study echolocation without a Click and an Echo.

Casey again exercised veto power. "Click" sounded too harsh, too mechanical to her ears. "Misha," she proposed instead. The dolphin had executed a high, graceful leap when we first saw him,

Casey contended. So what better sobriquet than Mikhail Barysh-nikov's nickname? Randy, inveterate scientist that he is, pointed out that there had been eight dolphins in the group at the time, so the odds were only one in eight that the dolphin we caught was actually the one who'd been leaping. Casey was convinced otherwise. And she managed to convince the rest of us. Misha it was. We'd named our two young male dolphins from Florida after a hapless Greek nymph and a Russian ballet star.

Whatever the dolphins thought of their recent experiences, it didn't seem to affect their appetites. Not only did they eat within half an hour of their arrival at LML, but by their second day there both Echo and Misha started hand feeding. Rather than just tossing fish to them as we'd done at Ruskin, we could hold the fish in the water and have the dolphins come take them from our hands. Echo tried it first, but Misha soon followed suit. Initially they were very tenta-tive, quickly snatching the fish from our hands and fleeing, swim-ming off to eat at a safe distance from us before coming back for more. Over the next few days they worked up to sticking around longer, taking as many as four or five fish before swimming off. They also learned to distinguish fish from fingers, for which we were grateful—even the gentlest of dolphins has a mouthful of sharp, pointy teeth (about eighty in a bottlenose dolphin).

We fed the dolphins from a wooden platform that extended over the water. We'd either kneel on the platform or lie on our stomachs, which put us face-to-face with the dolphins and allowed us to reach out to them better. Usually two people would feed, one for each dolphin, so we could give them more individual attention. To keep them from bumping into each other all the time, we as-signed them each a training station: Echo was supposed to go to the trainer on his left, Misha to the one on the right, though it took them a while to learn their designated sides. Michelle did the major share of the feeding and training, but with four or five feeds a day we all got our chance.

We used police-style whistles, which we wore around our necks, as a way of signaling the dolphins. Three toots on the whistle heralded the start of a feeding session. Within a session each whistle meant something like "right" or "yes" or "well done," and informed the dolphin that further reward—usually fish at that point—was forthcoming. If the dolphin did whatever we were hoping for (or engaged in some unsolicited, unexpected behavior we liked better), we'd blow the whistle and give him a fish. Initially we mostly hoped that they would eat from our hands, and do so gently; at this stage, then, we were essentially giving them fish to reward them for eating their fish.

Echo and Misha quickly learned to associate the sound of our whistles with food. Before long they weren't even waiting for us to blow the whistle to signify feeding time but would rush to their stations as soon as people appeared on deck carrying buckets. Within a week or so they even knew the feeding schedule and would admonish us, chuffing and slapping their tails on the water, if we were late with the fish.

As the two dolphins got calmer about feeding, we worked toward being able to touch them. Ultimately we wanted them to station with their snouts, or rostrums, resting in our hands during training sessions. We started by putting both hands in the water, one holding a fish, the other slightly in front of it. If the dolphin touched the foremost hand, we'd give him the fish. At first they apparently considered the extra hand merely an obstacle standing between them and their fish. Misha especially would bat at our hands, trying to get them out of the way. Gradually they got the idea, though, and became increasingly accepting of our touch. They grew accustomed to our hands, if not our faces.

Echo and Misha ate well, like the growing youngsters they

were. They each averaged about twenty pounds of fish per day during their first month at Long Lab. Initially they probably needed to increase their blubber layer from the thin summer coats they arrived with to something more suitable for our chillier waters. After a month or two their appetites slackened a bit: their daily totals thereafter generally ranged between twelve and seventeen pounds apiece, varying with time of year, their health, moods, and whatever. We adjusted the amount according to how well they were eating, increasing the poundage if they still seemed hungry at the end of feeds and decreasing it if they started to play with their food or act uninterested.

Their diet consisted of a combination of herring and capelin, supplemented by an assortment of vitamins. We bought regular over-the-counter vitamins—multiples, A, B12, C, and E—and stuffed the tablets down into the gill slits of the dolphins' first fish of the day. At first we fed them mostly the larger, higher-calorie herring, with just a few capelin for variety. As Echo and Misha progressed in their training, however, we gradually increased the proportion of capelin in their feeds. They would fill up in no time on the half-pound herring, whereas the capelin—each scarcely bigger than a frozen fish stick—gave us more to work with during training. That allowed us to make sessions longer and vary the size of the reward more—a nice, big herring or lots of little capelin tidbits for something done especially well.

We carried that principle one step further and after about a month started cutting a few of their herring in half. Echo and Misha were deeply offended. Dolphins generally swallow their fish whole, and these two clearly didn't want theirs any other way. Not only did they both spit out their first cut herring in apparent disgust, they also spat back several subsequent whole fish, evidently no longer trusting us to provide a decent meal.

We continued to try a few cut fish in their feeds periodically. Echo finally relented a month or so later. Misha held out several weeks longer, and even then he would eat only the head end; he was still spitting out tails in December. Ultimately both of them accepted even herring thirds and capelin halves.

We took turns going in for swims with Echo and Misha, one or sometimes two of us at a time, every other day or so. We generally limited our swims to about ten minutes and tried to give the dolphins as much say in the matter as possible. We didn't try to approach or touch them unbidden (not that we could do so, anyway—they could literally swim circles around us), instead letting them make the overtures. Mostly they kept their distance, at least at first. Now and then one or both of them would work up his nerve and swim by really fast, "buzzing" us with his echolocation. If they started chuffing or tailslapping, presumably indicating they'd had enough of us, we'd get out.

Or perhaps the chuffs were intended as chortles. I suspect the dolphins may have found our aquatic efforts more laughable than threatening. The sixty-five- or seventy-degree water was too cold for most of us thin-blooded Californians, so we wore wet suits. Without weight belts, however, the buoyancy of the neoprene suits made us bob about the tank like oversize bathtub toys, drifting ineffectually in the current created by the water-circulation system.

Holly probably was the only one the dolphins took at all seriously in the water. A true athlete—triathlete, actually—with the smoothly muscled physique of a competitive swimmer, she swam regularly in the gelid waters of Monterey Bay. Undaunted by the temperature in the dolphin tank, she was willing to forgo the wet suit. Holly didn't bob, she swam.

We would play with various toys in the tank—assorted balls, hoops, Frisbees, plastic baseball bats, and the like—trying to entice Echo and Misha into a game. Initially, they refused to play. They would watch us, however, and often one of them would take up whatever toy we'd just been playing with—though usually not until after all the humans had left the water.

Being a member of the dolphin crew didn't just mean feeding and swimming with the dolphins, however. Far from it. As Michelle liked to point out to starry-eyed prospective volunteers, "Dolphin work is ten percent glamour and ninety percent shit work."

Certainly lab attire was none too glamorous. We generally dressed in our oldest jeans and T-shirts or sweatshirts, since whatever we wore ended up stained with fish guts and full of holes, eaten into the fabric by salt water and the bleach we used to clean everything. Sometimes we wore heavy-duty yellow rain gear over our clothes, even when it wasn't raining, to keep from getting drenched during training sessions. Michelle practically lived in her "yellows." We could always tell she was coming, even at some distance, by the brisk swish-swish of her rubberized overalls.

We didn't smell very glamorous either. Eau de herring, with overtones of Clorox, was the prevailing scent. My friend (and now husband), Joel Shurkin, liked to describe the lab as the only place he knew where "you could see seven beautiful women dressed like street people and reeking of dead fish." We all had our stories: people backing away from us as we stood in line at the grocery store after a day at the lab; young children announcing to their mothers, and anyone else within earshot, "Mommy, that lady smells funny"; showering and getting all gussied up for a night on the town, only to find ourselves picking fish scales off our arms and faces at the dinner table.

During feeding sessions the dolphins sometimes exhaled right in our faces, spraying us with a somewhat sticky, salty mist. I usually ended up with my glasses so coated with goo, I could barely see. The gunk also had a way of giving your hair extra body. One night when Holly had to run straight from the lab to her job as a waitress without showering in between, several people commented on how nice her hair looked and asked what she'd used on it. "Dolphin snot," she replied.

We spent a good share of our day in the fish kitchen, where we stored and prepared the dolphins' food. It consisted of three rooms: a large walk-in freezer, stacked high with boxes of frozen fish; the "cold room," basically a large refrigerator, where we put the next day's boxes to thaw overnight and where we kept buckets of fish between feedings; and the "prep room," all operating-room white and stainless steel, where we fixed Echo and Misha's meals.

We sorted our way through great stacks of thawing fish, inspecting each one to make sure it was acceptable. We would eliminate any headless fish, of which there seemed to be many (plus the occasional UFO—unidentifiable frozen object). Toss out broken or torn fish, too, since even a small tear could provide a site for bacteria to grow. Poke the fish's belly to make sure it wasn't soft or mushy. Check for strange skin lesions or a bad case of freezer burn. Nothing but the finest for our dolphins.

We shared the fish kitchen with two other research groups, both studying sea lions, which were housed in smaller tanks near the dolphin complex. That meant the prep room was often jam-packed with people doing all manner of things to dead fish: inspecting them, cutting them into pieces, sticking vitamin pills into their gill slits, weighing out prescribed amounts and piling them into buckets, sometimes even pureeing them into "fish shakes" for young or ailing animals. We ground up the waste fish—as much as a quarter of what we'd thawed—in an industrial-strength garbage disposal, which Michelle and Casey dubbed Igor because of its ominous-sounding growls and snarls. If overloaded, Igor tended to spray fish gobs all over the room and everyone in it.

Disposal eruptions only exacerbated what was already one of everybody's least favorite chores: cleaning the fish kitchen at the end of the day. We washed everything—all the counters, sinks, shelves, walls, floors, doors, and windows in both the cold room and the prep room—with soapy water laced with Clorox, and then hosed down the whole place. Fish scales were the worst of it: they stick like burs and seem to multiply the harder you scrub, whether you're trying to get them off feed buckets, the walls and counters of the fish kitchen, or your own skin. Fortunately, the research groups took turns, so each crew got stuck with cleanup duty only one week out of three.

Unloading the shipments of fish that arrived every month or two also ranked high on the "shit work" list. Anybody who couldn't arrange to be elsewhere when the truck arrived would line up between the back of the truck and the freezer, and we'd pass each box from person to person in assembly-line fashion. The rawest deal, literally, was freezer duty: one or two people had to stand inside the freezer itself and pile the boxes onto pallets. On one such delivery I counted more than one hundred boxes, each weighing about forty pounds: I figured I'd held two tons of cold fish in my arms that day.

Rather less grueling but more exacting were the daily water tests. We'd take water samples from several locations around the tank and test chlorine levels, pH levels (an indication of how acidic the water was), and water temperature. We aimed at keeping the chlorine at no more than about 0.5 parts per million, lower even than Santa Cruz County's drinking water. If it got too high, the dolphins might start to look squinty-eyed and their skin would slough off more than usual. (A dolphin is always sloughing skin, at a much more rapid rate than we do. Some researchers have even suggested this may help reduce turbulence and drag as they move through the water, thereby increasing their swimming speed; in a sense, they may actually swim out of their skin.)

The chlorine by no means eliminated the algae problem, and we had to vacuum the dolphin tanks daily. The long-handled pool vacuum was unwieldy and the tank rim quite narrow in places; on more than one occasion the vacuumer, pushing the vacuum out to reach the center of the pool, followed the pole out and landed in the pool—to the consternation of both person and dolphins.

On warm, sunny days vacuuming was generally a pleasant enough task; we could watch the dolphins play, gaze out across Monterey Bay, and work on our tans while we were at it. It was less fun when the wind and fog chilled us to the bone. In cold winter downpours the task was especially onerous. Moving clumsily in our rain gear, we'd edge our way along the rain-slickened tank rim, barely able to see what we were doing, fingers too numb to hold the vacuum pole properly. Echo and Misha, impervious to the rain, may well have wondered why the humans seemed so grumpy.

The algae outpaced even our best vacuuming efforts, however, slowly turning the aquamarine tanks a deep mossy green. That gave rise to perhaps the least favorite job of all, even worse than cleaning the fish kitchen: cleaning the pools themselves. Every few weeks— more often during the spectacular summer algae blooms, less often in midwinter—we'd have to drain the water out of the tanks (one at a time) and scrub them by hand. The first step was to move the dolphins from the large pool into the thirty-foot pool (or vice versa) and put in a watertight gate to seal off the passageways between them. Then we'd open the drain to dump the water out, running it first through a dechlorinization system and then back out to sea. Once the pool was empty, we'd pour some thirty gallons of Clorox down the sides to burn off the algae. I suspect we set a record for bleach purchases at the local Long's drugstore.

After letting the bleach do its job awhile, we'd climb down into the pool with a battery of scrub brushes, brooms, doodle-bugs—soft, nonscratching pads for cleaning the windows—and a long hose. Dozens of years of college among us, and there we were, on our hands and knees scrubbing the floor.

The pool bottom slanted down toward the drain in the center, and the Clorox-sodden algae made it treacherous to walk on. We slipped and slid about as we worked, trying desperately to avoid the ominous slough of highly concentrated bleach water in the center, certain we'd melt on contact, like the Wicked Witch of the West in water. We probably did melt some brain cells while we were at it—the evaporating bleach produced a cloud of chlorine gas that hung about us as we worked, often leaving us with headaches.

In retrospect, we probably should have used those brain cells a bit more effectively to insist upon some viable alternative to the Clorox-inundation method. Finally, two years later, after several abortive attempts—including looking into outfitting ourselves with gas masks, which proved to be prohibitively expensive—someone finally hit upon the idea of using high-pressure hoses as a safer, saner pool-cleaning technique.

Visitors to the lab sometimes asked why the dolphin crew consisted entirely of women. I never came up with a satisfactory answer. Maybe women are more willing than men to work for little or no pay, to work simply for the love of the animals. Most of our crew members volunteered their time, and even those who got paid clearly weren't in it for the money. Or maybe women feel drawn to dolphins, like young girls to horses. Maybe we pick up on some feminine or "yin" quality to dolphins. Or maybe it was pure happenstance, just the luck of the draw.

Certainly several men were deeply involved in the project—Ken as "Grand Pooh-bah," Randy as project coordinator, Howard as vet tech—and much later we did take on a couple of male crew members. But on a day-to-day basis the dolphin project clearly was a predominantly female-run operation, so much so that Randy jokingly dubbed the crew the "dolphinettes." Given the various meanings of the suffix "ette," that would make us diminutive dolphins, substitute dolphins, female dolphins, or some combination thereof. So be it. After some initial umbrage at the appellation, we decided to claim the title as our own: we were the dolphinettes, and proud of it.

Echo and Misha underwent any number of changes in adjusting to their new environment. At first, as we discovered through periodic twenty-four-hour observations, they swam continuously throughout the day and the night. Some of the time they swam very slowly, apparently in resting mode, but they kept on swimming. That's probably typical wild-dolphin behavior—with sharks on the prowl, you can't afford to settle in for a good night's sleep. Dolphins also breathe voluntarily, unlike terrestrial mammals, so they need to maintain at least some degree of consciousness to continue breathing. (Whether and how dolphins actually sleep remains a matter of some contention.) By mid-August Echo and Misha had begun resting on the bottom of the tank, stretched out side by side on the concrete. They would rise briefly to the surface once or twice a minute to take a breath, then return to their roosts. The two of them seemed to take turns standing "guard duty," one rather more alert, the other apparently in a deeper state of rest.

They also started paying more attention to the humans in their environment. Increasingly, they'd stick their heads up out of the water to look at us when we came out onto the deck, watching us as we worked, even following us as we moved around the tank. At first they gave us mostly one-eyed stares, just tilting their heads slightly to one side. Gradually they did more "spyhopping" as well, poking their heads up entirely out of the water and looking around with both eyes. Dolphins and whales do spyhop in the wild, apparently to check out their surroundings visually. Captive dolphins may do so more frequently, however, since a larger share of the action of interest to them now occurs above the surface, where the people are.

They would also observe us as we observed them through the underwater viewing windows. What did they make of that invisible barrier? Of humans who appeared to be under water but weren't? Some sound passed through the windows, so they could hear us, especially if we tapped on the windows or clattered about with the equipment in the viewing room. But they couldn't echolocate on us. Presumably they could see three-dimensional figures moving about but would hear echoes only of a smooth, flat surface. Such an alien world, where even their senses didn't function properly.

All the equipment and contraptions we used must have baffled them as well. For the first three weeks Echo and Misha assiduously avoided the pool vacuum, retreating to the far side of the tank whenever we brought it out. Then one day, after they'd apparently adjusted enough to start feeling a bit bolder, to consider the pool their turf, perhaps, they started responding to the vacuum as though it were a hostile invader—some bizarre flat-headed beastie with a long sinuous tail, noisily scraping and sucking its way along the tank bottom. Both Echo and Misha squared off against the vacuum-monster, hovering in the water about ten feet from it, and threatened it with a series of jaw pops—loud popping noises dolphins apparently use in aggressive interactions.

The sounds are sometimes called jaw claps, because dolphins generally open and close their mouths in the process and so may seem to produce the sound by clapping their upper and lower jaws together. Some researchers suggest otherwise, however, and I'm inclined to agree. As I watched Echo and Misha fire away at the vacuum, I could see their throats bulge out, as though filling up with air. It reminded me somehow of the air rifle I had as a kid: first you cocked the handle to load it with air; then, when you pulled the trigger, the rifle made a loud *pop*. Dolphins, too, may be loading up some internal mechanism with air to make their pops, rather than merely clapping their jaws.

That first day Echo and Misha railed against the vacuum for forty-five minutes straight. Over the next week or two, the popping bouts got shorter and shorter, as the dolphins gradually seemed to accept the vacuum—whatever they thought it was—as a regular inhabitant of their new environment. Later still they became so accepting of it, they started treating the thing as a sex toy: they'd catch the vacuum hose in the crook between their bellies and their erect penises and swim off with it, often disconnecting the hose from the vacuum head and making our job nigh onto impossible.

Echo and Misha had to adapt not only to us and to the tank but also to each other. Even if they'd known each other beforehand, they still needed to readjust their relationship in accordance with their new circumstances. On the basis of their sizes, we estimated that Misha was about a year older than Echo—seven years old to Echo's six. Echo also seemed to act younger, at least at first. He whistled more. He seemed more easily frightened. He swam more often in a subordinate or "calf position" relative to Misha—close by his side but slightly behind him. Young calves tend to swim close to their mothers in this way, partly for safety purposes, but also to be carried along in the pressure wave created by the mother's swimming motion. It was probably this tendency that led dolphins to discover they could hitch a ride from larger whales; they ultimately took this finding a step further and learned to ride the bow waves of boats as well.

The two of them jostled and vied with each other somewhat for position during feeding sessions. Initially Misha seemed bolder, more aggressive about it. He generally came in more readily to feed and often tried to take fish from both trainers, rather than just staying on his designated side. He'd even push Echo out of the way, or angle himself at the platform so that his body effectively blocked Echo's access to the food. He usually finished first and then managed to steal some of Echo's fish.

If Echo and Misha kept their own "shit work" list, they would undoubtedly place medical exams and "gating" near the top. In order to monitor their health status, we gave them regular physical exams. They got their first full physicals on July 28, a week after they arrived, and then routine exams every three months or so thereafter. We'd move up the schedule, or sometimes just take a blood sample, if either of them evidenced any health problems.

Exam days tended to be a major undertaking, for dolphins and dolphinettes alike. To get them out of the water so they'd be more manageable, first we'd move Echo and Misha into the thirty-foot pool, seal off the gates to the other pools, and drain the water. Then we put them onto foam pads on the bottom of the tank, with six or seven people holding each animal. Both ends of a dolphin are potentially dangerous, should he decide to fight. Not only can they bite, but their rostrums are extremely hard and can hit you like a billy club if they swing their heads. The flukes, which propel a dolphin so forcefully through the water, can also propel people through the air, into walls.

We took samples of their blood, blowhole mucus, stomach contents, urine, and feces. We weighed them, measured lengths and girths all up and down their bodies, and gauged the thickness of their blubber layers using a portable ultrasound device.

Having all those humans handling them, poking, prodding, and sticking tubes into every orifice, must certainly have been unsettling and uncomfortable. I'm none too fond of physicals myself. But I suspect for Echo and Misha the worst of it, especially that first time, was to have their water—their world—literally disappearing down the drain. They were accustomed to water that rose and fell according to predictable tide schedules; water that could suddenly, inexplicably, drop out from under them—that was a different matter.

For veterinary exams, tank cleaning, and so on we needed to be able to move the dolphins from one pool to another. That meant "gating." We wanted them to swim through the gate into the adjoining pool, on cue, when we splashed a target (a plastic pool float attached to a wooden handle) on the water in the desired tank.

Echo and Misha seemed genuinely frightened at the prospect. Even though we'd left the gates open since their arrival, they'd never once swum through on their own. So much for curiosity. We tried luring them through with fish, to no avail. When we went in swimming with them, we made a point of going through the opening into the thirty-foot pool, hoping they would follow our lead, or at least decide it must be safe if the humans were doing it. They peered after the swimmer as she disappeared from sight but never made any attempt to follow.

In late August we started to work on gating in earnest. First we pulled a crowder net across the tank to force them into the corner near the gate and, we hoped, through the opening. Instead they both sank to the bottom of the pool. Howard, dressed in a wet suit (plus weight belt—bobbing wouldn't do at this point), got into the water to help guide the recalcitrant dolphins through the gate. Misha swam through, unassisted, into the thirty-foot pool. Echo escaped under the net into the main part of the large pool. We pulled the net out and started over. This time, with both Howard and Michelle in the water escorting him, Echo finally went through the gate—backward.

We gated them back and forth several more times that day. Misha swam through first each time; Echo continued to need more coaxing. We stopped for the day with them in the thirty-foot pool, hoping they would want to be back in the large pool badly enough to swim out again on their own. Not for love nor herring. They were still in the smaller tank when we arrived the next morning.

We continued regular gating practice over the next few weeks, having Echo and Misha swim back and forth between pools. They weren't happy about doing it even once, let alone several times in a row. The dolphins would work themselves up to swim through the gate one way, and then we wanted them to turn around and do it again the other way. Back and forth, back and forth. They looked at us as if to say, "Just make up your minds already, would you?"

Gradually, Echo and Misha got more used to it all. We didn't need people in the water after that first day. Soon we didn't even have to pull the net all the way across the pool—splashing one end of it into the water usually sufficed to let them know we were serious. Sometimes just banging the lid of the garbage pail in which the net was stored would do the trick. I felt a bit like a parent saying "You'd better be in that pool by the time I count to three or else . . ."

They got better about gating over time but never seemed fully comfortable with it, even after they'd done it many times. Usually they went through with just the splash of the target, no net. Often it took several target splashes, however, before they'd make up their minds and dash headlong through the opening—generally with both of them trying to go through at once, or as close together as possible.

We never knew for sure why gating seemed such a dreadful thing to Echo and Misha. Apparently they weren't alone in that, however; dolphin trainers elsewhere have reported similar problems with some of their animals. Ken Norris suggested that dolphins—at least coastal types like Echo and Misha, whose home environment may well include rocks and other natural barriers in the water—may be loath to enter any place they can neither see nor echolocate into. Sharks might be lurking just around the corner. Maybe the passageways between tanks even reminded Echo and Misha of the passes linking Tampa Bay and the Gulf of Mexico back home, which were indeed prime shark territory.

Once they'd been in and out of a tank several times in a row, though, you'd think the dolphins might be convinced there were no sharks there. Never at any time during their stay did they meet a single shark in the tanks at Long Marine Lab. Still, in the wild, just because there's no shark behind a rock one time you pass by doesn't necessarily mean there won't be one there the next time. And in captivity—well, who knows what the crazy humans might come up with next? Echo and Misha had no way of being certain.

We soon added to their uncertainty, I suspect. Before long they would indeed discover strangers—albeit not sharks—in the adjoining tank. The intensive gating practice, in part, had been in preparation for the arrival of "the Lags."

3

Strangers in the Tank

The strangers in the tank were dolphins, too, but a different kind from Echo and Misha—a species our Florida pair certainly had never encountered back home. The newcomers were smaller, quicker, and noisier than the two bottlenose, with shorter snouts and more sharply hooked dorsal fins. They were beautiful, intricately patterned in black and white and shades of gray. Their common name, Pacific white-sided dolphin, was almost as much of a mouthful as their full scientific name, *Lagenorhynchus obliquidens,* so we settled for a shortened version of the latter and called them simply "the Lags."

Long Marine Lab, and the tank complex Echo and Misha had thus far had to themselves, was to serve as a temporary holding facility for two groups of these dolphins. Miami Seaquarium and Chicago's Shedd Aquarium each wanted to catch four Lags for their exhibits. LML was ideally situated to assist in the undertaking.

Monterey Bay, with its deep submarine canyon and its upwelling of cold, nutrient-rich waters from the ocean bottom, is marvelously productive. In fact, it was recently named a national marine sanctuary. Among the many life-forms it supports, at least part of the year, are more than twenty species of cetaceans (whales, dolphins, and porpoises), from the common dolphin to rare beaked whales, from the roughly human-sized harbor porpoise, one of the smallest members of the clan, to the hundred-foot blue whale, greatest of all leviathans—not only the largest of the whales but the largest animal ever to have lived on this planet. The most abundant cetacean of Monterey Bay, however, is undoubtedly the Pacific white-sided dolphin.

The four Lags destined for Miami were to spend only about a month with us, an initial adjustment period to make sure they were eating properly and adapting well to captivity before they were transported across the country—similar to Echo and Misha's week-long stay in the bay pen at Ruskin. The Chicago-bound four would remain with us much longer (two and a half years, as it turned out, even longer than we'd bargained for), since Shedd was still building the tanks to house them.

Ken Norris had long sought the opportunity to observe Lags close-up and in depth; they were our next-door neighbors, and he wanted to get to know them better. Shedd wanted their animals well accustomed to captivity, well acquainted with training procedures, and ready to go as soon as their new aquarium opened. We could provide the pool and the people. Shedd, in turn, could provide the dolphins and the funding for the Lag study Ken wanted. Shedd's financial contribution would also help pay the considerable costs of keeping our dolphin facility running, making the arrangement all the more attractive, since our budget for the Echo and Misha project was already seriously strained.

For the crew, however, the coming of the Lags was something of a mixed blessing. The animals themselves were a joy; none of us could resist their grace, beauty, and ebullience. I learned more about Echo and Misha, too, I believe, by getting to know this other species, by having the Lags for comparison. On the other hand, these new animals took time—already scarce—away from our work with Echo and Misha, who themselves were still in the early stages of training; we hadn't yet begun the research work, and now that would be put off even longer. Furthermore, the Lags' arrival again brought to the fore all our questions and concerns about keeping dolphins in captivity, perhaps even more strongly than the capture of Echo and Misha.

Bottlenose dolphins, like Echo and Misha, are what most people in this country think of when you say "dolphin." That's what Flipper was (or were, rather—several dolphins played the role). That's what you'll most likely see on display in oceanariums, since they are relatively common along much of the U.S. coast, and they tend to fare well in captivity.

The bottlenose, however, is only one of some thirty or forty species of dolphins, out of roughly seventy-five species of cetaceans. The exact numbers depend on who's doing the counting. Taxonomists, the scientists who classify organisms into various categories according to how closely related they are, don't always concur in their classifications. They describe an animal's lineage in terms of the following major category types, in descending order of size and level of organization: kingdom, phylum, class, order, family, genus, and species. These may, in turn, be further divided or regrouped into suborders, superfamilies, subspecies, and the like. Human beings share a kingdom (Animalia) with all animals but are a species (*Homo sapiens*) all our own.

We would be lumped together with Echo and Misha, and with the Lags, down through the class level; dolphins, like humans, are mammals, members of the class Mammalia. We part ways when it comes to order: humans belong to the order Primates, along with apes and monkeys; dolphins belong to the order Cetacea, along with whales and porpoises.

The term "cetacean" derives from *cetus,* the Latin word for whale, so technically you could call all of them whales: Dolphins and porpoises are small whales, the larger variety are great whales. The distinction is more a matter of veneration than of science—if it's big enough, call it a whale. The killer whale is actually the largest member of the dolphin family.

The distinction between dolphins and porpoises does have some scientific basis, though even marine mammalogists will often use the terms interchangeably. Technically, the two refer to different families: the thirty or so species of oceanic dolphins—which include both the bottlenose and the Pacific white-sided—all belong to the family Delphinidæ, and the six species of true porpoises belong to the family Phocœnidæ. River dolphins are yet another story, with four or five species belonging to anywhere from one to four families, again depending on who's counting.

There are a few rule-of-thumb differences between phocœnids (porpoises) and delphinids (dolphins), though like all rules they have plenty of exceptions. Porpoises have blunt, somewhat rounded faces, without the elongate snout that distinguishes most—but not all—dolphins. Porpoise teeth are spade-shaped, whereas dolphin teeth are conical. Porpoises tend to have triangular dorsal fins, while dolphins generally have recurved, sickle-shaped dorsals. A couple of dolphin species have nearly triangular dorsal fins, however, a few have somewhat rounded dorsals, and another two have none at all. There's also a finless porpoise.

Confusing matters further is the dolphin fish (also known as mahimahi or dorado), which is not even in the same class as dolphins. If you see "dolphin" listed on a restaurant menu, it's the fish, not the mammal, at least if you're in the United States. Dolphins are protected here, though people do eat them in some countries.

The Lags are members of the same family as Echo and Misha, but they belong to a different genus. Their genus, *Lagenorhynchus,* has five other member species, all qualified to be called Lags. We often referred to Echo and Misha by genus as well, calling them "the Tursiops." The bottlenose dolphin's full scientific name is *Tursiops truncatus.* Like us, they are an "only species," sole members of their genus.

The family resemblance is considerable, though I doubt anyone would have much difficulty telling the two species apart just by looking at them. A bottlenose dolphin, as the name suggests, has a pronounced, cylindrical beak or rostrum. A Lag's rostrum is much shorter and broad-based, more triangular than bottle-shaped.

Bottlenose dolphins are also bigger. Echo and Misha might reasonably expect to reach eight to nine feet in length and weigh five or six hundred pounds or more as adults. At just over seven and a half feet and three hundred pounds, they were already larger than any Lag was likely to grow. These Lags, not quite fully grown, averaged about six feet long and two hundred pounds each.

Probably the most dramatic difference is in their color patterns. Bottlenose dolphins are gray. Intelligent, personable, graceful—and gray. Individuals vary, ranging from a pale, cloudy gray to a dark charcoal. They can even "tan," turning darker when they get more sun. Many have subtle lines or markings on their backs or faces, like Echo's backgammon-board forehead. Their bellies are a lighter shade and may turn quite pink when they are sexually stimulated or otherwise aroused. Basically, though, Tursiops are gray.

Lags are flashier, higher-contrast animals. They have a black back, light gray sides, and a pearl-white belly. A finely etched black line separates the gray from the white underside. The black of the back is interrupted by a pair of "suspenders," white stripes that begin on either side of the face, curve upward over the top of the head, and extend on past the dorsal fin, ending with a splash along each flank. Anomalously colored Lags do turn up quite often, however. One virtually all-white individual, sighted regularly in Monterey Bay, is known among local dolphin researchers as Moby Lag.

The first of the Lags arrived at LML in October of 1988, just three months after Echo and Misha. We were lucky, I guess, they didn't arrive even sooner. Ken had made the arrangements with Shedd back in March. The original plan had been to catch the Lags in Monterey Bay at roughly the same time that we caught the two Tursiops in Florida. The catch team gathered in July, but the wind and waves and fog were so bad the boat couldn't even go out on some days. When they did make it out, they couldn't find, much less catch, any Lags. The group reconvened in October and tried again, this time successfully.

Marine mammal veterinarian Jay Sweeney again headed the operation, acting as "collector of record," as he had with Echo and Misha. There were more dolphins to catch this time, and the capture period was protracted, so Jay did a lot of traveling back and forth between Santa Cruz and his home in San Diego over the next couple months. Scott Rutherford, a consultant to oceanariums and part-time professional dolphin catcher, worked with him. With his weathered good looks, lean, rangy frame, and a bit of a swagger in his walk, Scott looked like a cowboy. In a sense he was—a marine cowboy, riding and roping dolphins for a living.

The third main member of the team was Dawn Goley, like me a graduate student under Ken Norris's tutelage. She would carry out the behavioral observations Ken had in mind, making the Lags the subjects of her doctoral thesis. Dawn truly embodied the ambivalence all of us felt, to some degree or another, about the undertaking. For her the project was fraught with intense conflicts and highly charged emotions. Her broad, quick smile and joking manner overlay a deep sensitivity, and much as she yearned to study dolphins, she was also profoundly distressed by the prospect of bringing them into captivity. Her decision to be a party to it caused her no end of grief, some of which she passed on to those around her.

Officials from both Miami Seaquarium and Shedd also took part in some of the captures, as did Howard, Holly, and several other of our regular crew members. Others of us declined. I couldn't bring myself to help with the catching, though I did feed and work with the Lags once they were at the lab. Randy, staunchly opposed to removing dolphins permanently from the wild, even turned down possible funding through the Lag project.

The Lags were caught in quite a different manner from our two Tursiops. Although some bottlenose dolphins do form open-ocean schools, many inhabit bays and coastal waters and so can be encircled in nets in shallow water and then hand caught, as Echo and Misha were. Lags, however, tend to be offshore animals. Though they may at times enter the shallows, Lags in this area generally favor the seaward edge of the continental shelf, at depths clearly not amenable to such capture methods.

Lags also tend to congregate in much larger groups or schools than Tursiops. A solitary Lag would be virtually unheard of in the wild, whereas a bottlenose dolphin by itself isn't unusual in coastal areas, and pairs and trios are quite common. Lags may be found in groups as small as ten or fifteen or in great aggregations numbering in the hundreds, even thousands, including animals of both genders and all ages.

The trick in catching Lags or other deep-water dolphins is first to get them bow-riding. As sailors and other ocean travelers well know, many species of dolphins (and at least one species of porpoise) will often hitch a ride in the bow waves created by passing boats. Some scientists may argue that this is largely a matter of energetics—dolphins can travel faster, more efficiently, by taking advantage of the pressure wave generated by the boat's motion. Undoubtedly. But to say that's all there is to it is a bit like saying people ski or ice-skate simply for its energetic efficiency. Who says energetic efficiency can't be fun, too? And why should a sense of fun be an exclusively human province?

Tursiops are even known to bodysurf, riding breaking waves in toward shore, where presumably more practical matters wouldn't lead them. One surfer acquaintance claimed to have spent an entire afternoon surfing alongside a half-dozen or so bottlenose dolphins just south of Santa Cruz. I've seen dolphins change their course entirely and come from quite a distance for a chance to ride the bow—they're not necessarily just catching a boat that happened to be passing their way. Nor do they ride along sedately like some newspaper-reading subway commuter; they leap and gambol, criss-crossing daringly within inches of the front of the boat, seeming to egg each other on to greater feats of derring-do. Bow-riding dolphins have never failed to elicit shrieks and squeals of delight from the people on every whale-watching cruise or class boat trip I've been on. As Herman Melville put it (erroneously referring to dolphins as fish): "If you yourself can withstand three cheers at beholding these vivacious fish, then heaven help ye; the spirit of godly gamesomeness is not in ye."

But somehow that makes it all the more gut-wrenching to catch dolphins that way. You don't use food as bait, as you might with some other animals, but rather entice them with a good time, invite them to come out and play and then shanghai them, turning their own "spirit of godly gamesomeness" against them.

The Lag captures were a smaller-scale operation than Echo and Misha's, involving just one boat with only four to six people aboard. The boat, a twenty-five-foot Boston whaler, was called the *Nai'a*, the Hawaiian word for "dolphin."

Rather than encircling a group of dolphins with a huge seine net as we'd done in Florida, this procedure goes after one individual dolphin at a time using what's called a "hoop net." The hoop net consists of a bag of netting, with a purse string at one end, attached to a stainless-steel hoop. As a dolphin swims into it, the net detaches from the hoop, the purse string draws closed, and the dolphin is caught in the bag. The net is attached to a line, which in turn is tied to the boat. The dolphin can still swim on its own and can rise to the surface to breathe.

Scott, the dolphin cowboy, did the actual catching. He'd stand, net in hand, on a metal pulpit that extends from the bow of the boat. Once the crew found a school of Lags and got them to ride the bow, Scott would try to place the hoop directly in front of a dolphin as it surfaced alongside the boat. When he caught an animal, two or three people in wet suits would jump into the water and put the dolphin into a stretcher. Then they'd hoist the dolphin aboard.

Once the dolphin was on board, Jay checked its size, gender, and general health status. On a couple of occasions the newly captured dolphin seemed too agitated to be transported safely and so was immediately released.

Both Miami Seaquarium and Shedd wanted youngsters, like Echo and Misha, old enough to be independent of their mothers but not yet sexually mature. Both facilities also wanted three females and one male, since they hoped ultimately to breed the animals. Given four animals, that's the split likeliest to produce the most offspring—one male should be able to impregnate all three females readily enough. We dolphinettes did wonder, though, what if you were one of the females and the only guy in town happened to be hopelessly unappealing? Are there dolphin dorks?

The procedures for the trip back to the lab were similar to those we'd followed with Echo and Misha—monitoring respiration and heart rates, keeping the animal wet, putting A and D ointment on its back, and so on. In this case, though, rather than riding back on foam pads on the bottom of the boat, the animal was placed in an open fiberglass box, shaped to fit the contours of a dolphin—relatively narrow around the rostrum, widening for the head and chest, narrowing again along the peduncle (tail stock), and then flaring out at the end for the flukes. These transporters, which Jay had designed, were an improvement over the ones we'd used with Echo and Misha, and with Jo and Arrow before them. Again the animal's weight was partially supported by the stretcher, but this time the majority of the support came from water rather than from foam. A deep trough accommodated the pectoral fins, and the more fitted, dolphin-shaped design made it possible to float the animal using a relatively small amount of water. One of the old-style transporters full of water would have been utterly unwieldy, hardly a transportable transporter.

Only two transporter boxes would fit on the boat at a time, so the crew could catch no more than two dolphins on any given trip out onto the bay. Their first day out, the crew caught one Lag but ended up letting it go again when they were unable to catch another from the same group. They didn't want to go home with just one to start with, to have one lone Lag in the tank. For an animal that has spent its entire life as part of a group, always surrounded by others of its kind, being alone could perhaps be the most traumatic, the most alien concept of all.

The first successful catch—two young females—came on Monday, October 17. The boat landed at the Santa Cruz harbor about noon. The crew placed the dolphins, in their fiberglass boxes, in the back of a pickup truck for the twenty-minute drive to Long Lab.

The catch team had called ahead from the harbor to tell us to prepare for the arrival of the two Lags. That meant first moving Echo and Misha into the thirty-foot pool—no more practice, this was gating for real. Then we drained water from the large pool until it was about waist-deep.

Once the truck arrived, we hoisted the dolphins out of their boxes and carried them to the tank. After we lowered them into the water, Jay walked one of the dolphins around the tank, holding her in his arms and showing the animal the boundaries of her new home, as he'd walked Echo and Misha around the pen in Ruskin three months before. Scott walked the other dolphin in similar fashion. The rest of us, dressed in wet suits, lined up around the inside of the tank. Our job was to keep the dolphins from hitting the sides—by splashing our hands on the water to warn them away, and by putting our bodies between them and the wall. These were deep-water animals, unversed in walls. At least Echo and Misha had already experienced seawalls, natural barriers, narrow passes; by and large, these Lags knew only open water.

Neither Lag hit the wall. We stood in the tank an hour or more until Jay signaled us, a few at a time, to get out. Finally, when the two dolphins seemed calm enough and all the people were out of the pool, we raised the water level.

For the first day after these two Lags (and each subsequent one) arrived, we kept an around-the-clock watch on them, with at least two people on duty at a time, on rotating four-hour shifts. The two A.M. watch took real dedication—not only would you blow the whole night's sleep, you couldn't even see much of the animals while you were there, though we did keep a light on over the tank. I drew the six A.M. shift and watched the dawning of the first day of the dolphins' new life. The two Lags circled the pool slowly, swimming synchronously.

We weren't allowed to name these dolphins; that would be the prerogative of the powers-that-be at Miami Seaquarium and at Shedd Aquarium. For the time being we could only assign them numbers in lieu of names. The numerals quickly took on the full warmth and weight of names to us, however. By the time the Shedd people finally came up with official names for their animals—well over a year later—some of us had a hard time making the switch. Dawn never even tried; she still lovingly refers to the Lags by number.

So far we had One and Two. Three, another female, was brought in the following day. (Once we already had dolphins in the tank, bringing in one at a time was deemed acceptable.) We followed the same procedures as the day before, lowering the water level and walking the newcomer around the tank, with people posted all around as deterrents. Again we did a twenty-four-hour watch. Again I took the six A.M. shift. Now there were three, swimming in perfect synchrony.

Unfortunately, the Lags didn't start eating as readily as Echo and Misha had. The three new dolphins refused all fish we tossed into the tank, herring and capelin alike. We even tried squid, a major constituent of their diet in the wild. Not a nibble. When they still weren't eating by Thursday, Jay decided we needed to force the issue. At five that afternoon we dropped the water level down again, suited up in our wet suits, and climbed back into the tank. We used a crowder net to block off a portion of the tank, then circled the dolphins and caught them one at a time.

Several people held each Lag while Jay put a tube down into the animal's stomach and poured in water plus vitamins. Dolphins have no gag reflex, and they normally swallow whole fish much greater in diameter than the tube, so that wasn't quite as hard on them as it may sound. Dolphins get most of their fluid through the fish they eat, so dehydration was the most immediate concern. We would worry about getting solid food into them later.

That night the watertight gate separating the Lags from Echo and Misha popped open. The gates, made of plywood coated with fiberglass, were buoyant. They fit into grooves in the side and bottom of the channel between the two pools but were held in place by a difference in water pressure between the two pools: the water level in one pool always had to be higher than in the other, forcing the gasketing that lined the low-water side of the gate against the side of the groove and forming a watertight seal. Otherwise, the gate simply floated. To set a gate in position, two or three people (two large, three small) first had to stand on top of it to weigh it down into the grooves. That could be a bit of a balancing act, since the gates were only three or four inches wide; more than one person fell into the tank in the process. Then we raised or lowered the water in one of the tanks until the pressure differential was sufficient to hold the gate in place.

If we wanted to remove a gate, we equalized the water levels, again by either adding water to one tank or draining it from the other until the gate popped up—quickly and forcefully enough to clobber anyone standing in the way. It would bob in the water, still held upright in its grooves but with about a three-foot clearance beneath it, until we lifted it out. (Not as easy as it sounds—the gates were awkward, heavy, and usually slimy with algae.) Sometimes the gates would pop when we didn't want them to if the drain in one pool had been left slightly ajar, or if the freshwater inflow was a little too heavy in the other.

This was one of those times. Sometime during the night the gate popped up. We were no longer doing twenty-four-hour watches, so no one was tankside at the time. Echo and Misha, either taking the popping of the gate as a signal—after all the gating practice we'd done—or just tired of being in the thirty-foot pool, shot out underneath the bobbing gate and into the large pool with the three Lags.

Scott had been sleeping in the dolphin crew's office, a trailer overlooking the tanks, on call in case anything should go wrong. Shortly before dawn he discovered what had happened and, with the aid of the lab's resident caretakers, managed to get Echo and Misha back into the smaller tank and reset the gate without further mishap. Misha apparently grazed the bottom of the gate on his way out and came away with a couple of scrapes, one on the front of his dorsal fin, another on the lower tip of his rostrum. The dolphins were otherwise none the worse for their encounter.

We all arrived early the next morning, since we were scheduled to force-feed the Lags again at eight A.M. Scott described what had happened and suggested that we should deliberately let Echo and

Misha into the big tank and feed them in front of the Lags—maybe the Lags would get the idea that way. Dolphins are notorious mimics. In fact, they're the only animals other than humans known to be good at mimicking both sounds and movements: parrots and mynah birds are skilled vocal mimics; apes and monkeys are adroit physical mimics; humans and dolphins do both. Many trainers can tell tales of one dolphin learning a new trick just by watching another dolphin do it. If the Lags could just watch other dolphins —even dolphins of a different species—eat the fish we gave them, they just might catch on.

Dawn was horrified when she heard the plan. The two sets of dolphins could get into a horrendous fight, she insisted, and there would be nothing we could do to stop it. To an extent she was right. You think a dogfight is hard to break up, try separating several two- or three-hundred-pound animals in the water. If it came to that, we would need to start draining the tank.

Dawn argued. She pleaded. She cried. She accused the rest of us of not caring about the Lags. She told Michelle to go to hell and then fled to the underwater viewing room. I went after her. I was angry. I'd done my share of middle-of-the-night watches, spent long hours standing up to my waist in cold water, wrestled with my own conscience about taking part in this venture, shed tears of my own at seeing the Lags not eat. I cared. So did everyone else there. In fact, you'd be hard-pressed to find a group of people anywhere who cared more about dolphins generally and these dolphins in particular. I tried to tell that to Dawn, but she was in no mood to listen. I left, both of us still upset.

Dawn was outvoted. Given that the Lags and the Tursiops had been together the night before without incident, Scott's plan seemed worth a try.

We popped the gate, this time deliberately, and let Echo and Misha into the big pool with the Lags, calling them immediately to the side to be fed. The two Tursiops came right over and ate just fine. They and the Lags virtually ignored each other, showing little or no interaction of any kind, and certainly no aggression. No mimicry, either—the Lags still didn't eat.

Much later, after Dawn and I had a chance to become good friends, I learned more about what she called "one of the worst days of my life." The very night the gate popped accidentally, she'd been awakened by a ghastly, vivid dream: the two Tursiops got into the tank with the Lags and killed them all. When she arrived the next morning and heard we were planning to let Echo and Misha in with the Lags, she saw her nightmare about to come true. To make matters worse, what she'd understood as the rationale for trying it was quite different from the feeding-mimicry notion I'd been told. The idea, she thought, was to "jump-start" the Lags, essentially to scare them out of their wits to get them going.

Dawn is a field biologist, not a laboratory scientist, at heart; she prefers meeting animals on their terms, in their natural habitats. She had studied wild barn swallows for her master's degree, observed humpback whale mothers and calves off Hawaii, and worked with Randy on a study of blue whales in Monterey Bay. She is an outdoors person—a world-class rower, a rock climber, a runner— unhappy being physically confined herself, most unhappy about imposing such confinement on other creatures.

Her feelings were compounded by guilt: she would earn her degree out of the Lags' captivity. "Oh, God, if it weren't for me, these animals would be free," she lamented. That wasn't true, of course. Miami Seaquarium and Shedd Aquarium had their permits and would catch their Lags regardless of whether Dawn, or anyone from LML, were involved in the project. I knew the feeling well, however. And perhaps with more justification—I had more say in Echo and Misha's capture than Dawn had in the Lags'. But I had more faith than she did, I think, in the value of working with captive animals. I could also console myself with the knowledge that Echo and Misha were slated ultimately to return home, to freedom; Dawn had no such consolation.

We continued to force-feed the Lags two or three times a day, though after that first day we gave them herring rather than fluids through a tube. Initially that meant prying their mouths open, putting a fish in, and then holding their mouths closed until they swallowed—a bit like giving a pill to a dog. Gradually it became less a matter of force and more a matter of simply holding fish in front of them, or maybe tapping them lightly on the rostrum with the fish to get them to open up first. We still had to get into the water and catch each of the animals, however. It wasn't much fun from our standpoint—squeezing into still-soggy wet suits and jumping into cold water at eight A.M. to restrain a reluctant dolphin can get pretty old after a few days. I imagine it was even less fun from the dolphins' standpoint.

Lag Two swallowed a herring on her own that Friday, the day after they'd first been tube fed. One and Three held out a couple days longer, but by the middle of the next week all of them were eating regularly, without any coercion whatsoever. The catch team prepared to go out again.

The next two Lags—Four, the first male of the lot, and Five, another female—were caught on Wednesday, October 26. Six, also female, followed the next day. Again we dropped the water level down and introduced each new animal to the tank. We conducted twenty-four-hour watches after each arrival. Things got easier, at least in some respects, as the number of dolphins in the tank increased. With the first three Lags eating, the newer additions seemed to accept the fish more readily. Four, a feisty, energetic little guy, ate right away, his very first day in the tank. The new females were less eager but were eating on their own within a few days. In a little over a week Five was hand-feeding from the platform, spyhopping, and soliciting rubs from people. Six became the most aggressive eater of them all, even stealing fish from the others.

Other complications arose, however. Jay had arranged for ultrasound tests on all the Lags, and we discovered that Three was pregnant. We couldn't and wouldn't keep a pregnant female and so began to make plans to return her to the ocean.

Two presented another sort of problem. About a week after she arrived, she started bumping into walls and into other animals. She no longer kept pace with her companions, swimming more slowly, sometimes listing, often by herself rather than a part of the group. She continued to eat well, however, if we held the fish right in front of her.

Two had kept her eyes closed during capture and almost continuously thereafter. When we finally got her to open them, we discovered a whitish opacity in her right eye, a scarring of the cornea that apparently antedated her capture. She appeared to be completely blind in that eye. We couldn't be certain how well she saw from the left eye because she kept it closed nearly all the time. Presumably she saw well enough to survive in the wild; or she got by on her wits and her echolocation, and by being part of a larger school. In captivity she seemed to be having a harder time coping.

We faced a moral dilemma: since we'd caught her, did we owe it to her to do whatever we could, limited as that might be, to help her medically; or would she be better off back home, in the open ocean, with her own kind? If captivity had adversely affected her, should we treat her or simply end her captivity? Jay Sweeney, who knew better than anybody what we could and couldn't do for her, decided it would be best to free her.

The decision was the subject of some controversy among the crew. Some felt that letting her go again was a breach of our obligation to care for her, an obligation we'd taken on by catching her. This way we would never know if she lived or died; this way we wouldn't see if she suffered. Others were glad she'd be able to live free, and die free, whenever her time might come. I leaned toward the latter view, though I'll confess I was most happy not to have to make the call myself. Like so many issues in this business, it was a question that offered no clear-cut answer, and no way of knowing if you'd made the right choice or not.

On November 2 we loaded Two and Three back into transporter boxes and drove them to the harbor. This time the boat went out with two dolphins and returned with one. The crew found a large school of Lags and lowered Two and Three into the water among them. They both took off like shots. Blind or no, Two joined the others, swimming far better than she had in the tank.

While they were out there, the crew managed to catch yet another dolphin—Seven, a female. They went out again the next day and caught Eight, also female. A few weeks later they brought in two more females, Nine and Ten.

The group of Lags settled in. All were eating well. Most were even taking fish from our hands. Lags have razor-sharp teeth, however, and we all suffered numerous lacerations—like big, deep paper cuts—on our fingers and hands until the dolphins learned to take their fish gently and calmly. By the end of November the Seaquarium people felt their four animals were sufficiently adjusted and ready for the trip to Miami.

Not all was to go smoothly, however. Eight, upon ultrasound examination, had also turned out to be pregnant. On November 28 Scott, Dawn, and the others took her back out in the bay and released her into a school of some one hundred fifty Lags. She disappeared into the crowd.

Six had developed a small abscess on her flukes. Dr. Dave Casper treated the lesion after consulting via telephone with Jay Sweeney, who was now back in San Diego. Dave, a Santa Cruz veterinarian (known affectionately as Casper the Friendly Vet) served as Jay's local liaison. The wound was healing nicely, but the dolphin was on medication and still wasn't eating up to par, so it seemed best not to move her. Miami Seaquarium arranged a trade: they would take Seven instead of Six, who would stay at Long Lab and become part of the Shedd group.

On November 29 we all gathered at the lab in the early morning, loaded the four Florida-bound Lags—One, Four, Five, and Seven—into their transporters, set them in the back of two rental trucks, and watched with mixed emotions as the trucks pulled out of the driveway. We'd known them only about a month, but they had so filled our days and our psyches during that brief span, we couldn't help but miss them. Dawn cried, both, I think, because they were leaving—she was especially fond of Four, the sassy little male—and because they weren't going home. I was both sad to see them go and relieved that maybe we could start focusing more on Echo and Misha again. The Seaquarium trainer called later to let us know that the Lags had arrived safely in Miami and were settling in nicely.

The following day the capture crew went out and caught one more animal, Eleven, a male. Shedd now had its full complement of three females (Six, Nine, and Ten) and a male. Or so we thought.

On December 1, during routine physical exams on all the dolphins, we discovered another abscess, this one golf ball–sized, on Six's right flank. Jay Sweeney, who had returned for the exams, treated the new wound and prescribed more antibiotics. Both lesions were healing well, and Six became more active and playful over the next few days. Her appetite improved again.

On December 8 she quit eating; we had to force-feed her to make sure she got her medications. The abscess wounds looked fine—they weren't the problem. The next morning she started to vomit blood and fluid. Her respiration rate, normally about three or four per minute, increased to ten per minute. At twelve-thirty she started bumping into walls. Randy decided to move her into the twenty-five-foot tank, with a person in the pool walking her around. He called Jay for further instructions. As Randy spoke on the phone, Six began flexing her flukes high into the air. Her breathing rate shot up to twenty per minute. Just before three P.M. she died.

I'd been at class that afternoon. I got the word of Six's death by telephone, from Holly. There was to be a dolphin-crew Christmas party at her house that night, and she wanted everyone to know beforehand. We went ahead with the party. We needed it more than ever—to help heal some of the rifts between group members, as well as to provide each other some much-needed comfort.

In a subsequent necropsy Jay determined the cause of death to be acute bronchopneumonia, accompanied by gastritis. What the necropsy couldn't tell us was whether Six would have become sick and died had she still been back home in the bay. Pneumonia is a common problem among dolphins, however, both captive and wild. As Dave Casper pointed out, different species tend to have their own particular weaknesses with respect to health. We humans, for example, are most likely to die of heart disease or cancer. For dolphins, diving animals whose respiratory systems undergo constant stress, the lungs are their Achilles' heel.

Three days after Six died, the catch boat went out one last time and brought in Twelve, a small female. That was it. Shedd now had its four Lags, and the capture phase was over at last. It had been a long two months.

I suspect it had been a long two months for Echo and Misha as well. They spent those two months—and the next three, as things turned out—confined to the thirty-foot pool. We left the channel open between the two smaller, circular pools, so in theory Echo and Misha had access to them both. They never voluntarily swam through into the twenty-five-foot pool, however, so in practice they were left with just the one thirty-five-thousand-gallon tank.

The Lags had sole possession of the large tank during their arrival and adjustment period. For one thing, we were afraid they would be more likely to hurt themselves in the smaller pool. Not only are Lags accustomed to much deeper and broader expanses of water, they also tend to be even more energetic and acrobatic than the Tursiops. Lags surely would fit Melville's description of dolphin schools as "hilarious shoals, which upon the sea keep tossing themselves to heaven like caps in a Fourth-of-July crowd." Not so hilarious if they toss themselves to heaven and land on a concrete deck. We placed netting around the outside of the tank to help contain them.

We also had to consider the potential risk of the dolphins' hurting one another if they were in the same tank, something Dawn still feared, although different species of cetaceans have been kept together successfully at any number of facilities. Actually, it's hard to know just who might be at greater risk. Given their greater speed and agility, the Lags could well harass the Tursiops even more than vice versa.

Furthermore, Dawn planned to study the Lags' social interactions with one another, and she wanted those interactions as natural as they could be, given that the dolphins were "living in a bathtub," as she tended to put it. Insofar as possible, she wanted them socializing only with each other. Not with humans. Not with some other kind of dolphin, one they would never pal around with in the wild. She also didn't want to study trained dolphins—"people in dolphin suits," which is how she saw most of the show-trained variety.

That meant virtually no training with the Lags for the first year. Once they were all eating well, we just handed the Lags their fish at mealtime and otherwise left them largely to their own devices. It also meant separate tanks for the two sets of dolphins. The morning we let Echo and Misha in with the first three Lags to try to induce them to eat was the last time the bottlenose and the white-sided dolphins were ever in the same tank at the same time.

The goal ultimately was to play "musical pools" with the dolphins, moving them around and giving everybody equal time in the large tank. To avoid getting both sets of dolphins in the same pool at once, we first had to move whoever was in the thirty-foot pool into the twenty-five-foot pool, closing that gate off with a net; then the animals from the large tank went to the thirty-foot pool; finally the first set of animals could be released from the twenty-five-foot pool into the large tank. For the time being, though, we couldn't even train the Lags for gating.

Meanwhile, Echo and Misha were relegated to a pool thirty feet in diameter and six feet deep. At least they were well accustomed to depths of only a few feet back home, whereas such shallow water was a more alien experience for the Lags. The pool exceeded the size requirements specified by law for housing two bottlenose dolphins on a long-term basis. But that doesn't tell us what size requirements the two dolphins might specify. The tank seemed to look smaller and smaller as Echo and Misha's stay in it lengthened. My biggest regret of the whole venture, I think, was that I didn't press harder to get them back into the big pool sooner.

This business of catching and keeping wild dolphins took its toll on all the dolphinettes. The prospect of the Lags' permanent captivity, Six's death and the controversy regarding Two's release, and now Echo and Misha's lengthy confinement in the thirty-foot pool, all weighed heavily on us. Two crew members quit as a result, in part, of their feelings about these issues. Casey left, too, at the end of December, though for unrelated reasons.

Echo and Misha didn't stretch their flukes in the large tank again until the end of March.

4

Dolphinalities

*A*lmost from the outset we began referring to Echo and Misha as "the Boys," a carryover, in part, from having called Josephine and Arrow "the Girls." At least in Echo and Misha's case the label was age appropriate. They were youngsters still, just verging on adolescence. Jo and Arrow were another story. Arrow, around twenty years old, was fully grown and a mother. Her son, somewhat older than Echo and Misha, was still with the Navy back in Hawaii. Jo we estimated at "thirtysomething," well into dolphin middle age.

By referring to two mature females as "girls," I suppose we committed a breach of feminist consciousness. But then, we called ourselves "dolphinettes," and we weren't exactly youngsters, either. More than half of us who had worked with Jo and Arrow—including Michelle, Casey, Jill, and me—were in our thirties, as were honorary dolphinettes Randy and Howard. Holly was close. Another crew member was in her fifties. At that point only Kim and two others were under age twenty-five. We were a crew of "old girl" dolphinettes to match those "old girl" dolphins.

Echo and Misha were "young boy" dolphins, and the composition of the crew seemed to change correspondingly, in age if not in gender. One by one, and for a variety of reasons, nearly all the original Jo-and-Arrow crew left, to be replaced by a younger crowd. Eventually, most of the people who worked with the Boys (and with the Lags) were college "kids" themselves. Females continued to predominate on the crew—two or three males joined much later on—but the average age dropped about ten years.

By referring to two dolphins, whatever the age, as "boys" and "girls," we were undoubtedly guilty of anthropomorphism—a tendency to attribute human motivations or characteristics to animals, inanimate objects, and natural phenomena. It's what we do when we name a hurricane Bob, talk about a boat as if "she" were a lover, or put words in a dolphin's mouth. It's one of the seven deadly sins in science. It's one I've had to struggle with in writing this book.

We're perhaps most prone to anthropomorphize creatures whose resemblance to us we find flattering. Dolphins seem especially cute and clever, so we may be doubly tempted to speak of them in human terms.

It's hard to avoid a humanized vocabulary when talking about such qualities as sentience, intelligence, emotion, ethics, aesthetics, language, humor, courage, character, soul. "Personality" even has the person built right into it. But I can't believe we're so very different from the rest of the animal kingdom that none of the vocabulary applies. Clearly we share certain features of anatomy with other animals. The bones of a cetacean's flippers, for example, are homologous—that is, they correspond in structure and evolutionary origin—to those in the human arm and hand. Why not homologies in other regards, too? We are not, in my opinion, the sole sentient species on this planet, nor the only one with character. Nor is it simply a matter of adding dolphins to the list. Even our most human of attributes can be found, to some degree or in some form, across a wide range of species.

If I sense what another person is feeling, it's called empathy; if I do the same with an animal, it's anthropomorphism. I know I'm guilty of "Carolmorphism" with respect to other people. I'll attribute certain feelings or motivations to someone else to the extent their behavior is somehow "Carol-like." Often enough I'm wrong, and I have to be willing to change my attributions as I get to know the person better. But how else can I empathize, except by using my own feelings and motivations as a reference point?

Empathy may be as valuable in getting to know another species as in getting to know another person. But just as I'm likely to interpret my emotional responses more accurately the more I know about another person's temperament, gender, age, religion, ethnic or cultural heritage, and so on, my interpretations concerning another species should also improve the more I learn about their biology, ecology, and behavior. The trick, I think, is not to mistake the empathetic understanding for a scientific understanding, but rather to allow the two to complement and inspire one another. That has been my aim throughout this book.

Dolphins exhibit individual as well as species differences in both looks and "character." *Tursiops* may be among the blandest of dolphins in appearance, but they are perhaps more individualistic than most in temperament.

Bottlenose dolphins can look so uniformly gray to the untrained observer that visitors to the lab often wondered, "How do you tell them apart?" The dorsal fin (the one on the back) provides the primary means of identifying individual dolphins in the wild. The fin's overall shape and size, together with its pattern of nicks,

notches, scars, and other markings, very often are distinct and rec-
ognizable.

A dolphin's dorsal fin tends to take a beating over the years,
getting clipped or torn or gouged as a result of skirmishes with
fellow dolphins, with their physical environment, or with the human elements of their environment. Part of a fin may be sheared off
in a close encounter with a boat propeller. Fishing nets or even a
casually discarded piece of twine that gets wrapped around the fin
can slowly slice through it. One way or another, few dolphins make
it past childhood with a clean, unblemished dorsal fin.

Randy Wells and his colleagues have identified more than six-
teen hundred dolphins in the Sarasota–Tampa Bay area and nearby
Charlotte Harbor, based largely on photographs of the animals'
dorsal fins. They divide the photos into several categories depending on which portion of the fin has the most distinctive marking—
leading edge or trailing edge, tip, middle, or base, and so forth—
and on the nature of the marking. This gives rise to such precisely
descriptive names as Mid Scoop Low Box (the dolphin has a scoop
out of the midportion of the trailing edge of its dorsal, plus a box-
shaped chunk missing in the lower portion of the fin) or Nick Low
Mid Notch (a nick in the lower dorsal, plus a notch in the middle
third). Some more fanciful names are based on the overall shape of
the fin—Whistler's Mother (the fin is shaped like a silhouette of the
famous painting), Smurf (the dorsal resembles one of the cartoon
characters' hats), or Nose Picker (don't ask).

Other parts of the dolphin besides the dorsal can form the basis
of identification as well—prominent scars on other parts of the
body, some unusual coloration, a misshapen rostrum, and so forth.
Bubblegum was so named because of a pink blob-shaped patch of
skin, perhaps a birthmark. Wee Willie, apparently a dolphin
dwarf—so small he stayed with his mother until age ten—can be
identified by his diminutive size and somewhat misproportioned
body. Some less-distinguished-looking dolphins are identified only
by number. Even then, however, as with the Lags, for Randy and

crew these numbers have taken on the full warmth and weight of names.

Echo and Misha both have fairly distinctive dorsal fins. Somewhere along the line, well before we met him, Echo had lost the very tip of his dorsal fin, so it's slightly squared off at the top. He also has a few small nicks along the trailing edge. Misha has several notches along the trailing edge of his dorsal, including a larger, rounded one in the midportion.

Randy thought the rounded one looked like the kind of notch left by a small identification tag after it has worked its way out. He looked through the photo records of animals that had been caught and tagged in Tampa Bay and, sure enough, found a match. In late 1984 Randy had accompanied commercial collectors who were catching dolphins for display facilities; he went along to measure and tag the dolphins that were caught but not chosen by the collector, and to obtain blood samples for genetic studies. On one such trip Randy had tagged a three-year-old male—part of a group of juveniles just offshore from the Bahia Beach Resort—whose dorsal shape was an exact match for Misha's: you could lay one slide on top of the other and they would line up. The notch in Misha's fin was precisely where the tag had been placed. So we knew Misha had been three years old in 1984 and living in Tampa Bay, not far from where we caught him four years later.

The dorsal fin is often all you can see of a dolphin in the wild, so it's fortunate they tend to be distinctive enough for individual identifications. With captive dolphins, however, we can learn to recognize their individuality on other bases as well. We get to know their faces, the way they move, their mannerisms.

Echo has a little glitch on the left side of his mouth, about two thirds of the way back from the tip of the rostrum, which turns that built-in dolphin smile into something slightly skewed.

Misha has bigger eyes—except when he's sick; then he gets all squinty-eyed.

Echo and Misha have different patterns of coloration on their melons, the fatty portion of the forehead. Echo has his backgammon-board pattern, with elongate triangular tongues of dark gray alternating with a lighter gray. A narrow, very pale gray stripe begins at his rostrum, arches up over each eye like white eye liner, and trails off to his pectoral fins. Misha has his own more subtle patterns of light and dark on his forehead. His melon is rounded, more bulbous than Echo's.

Some of these differences are undoubtedly hereditary, leading to family resemblances. Certainly nicks and notches on the dorsal aren't, but the overall shape of the fin may well be, as may melon shape and pattern. The Dolphin Research Center in Florida (where Jo is now) has three generations of a family that all show the same sort of straight, unrounded melon. Owner Jane Rodriguez calls them the Flatheads.

You also have to be able to tell the "boys" from the "girls," not necessarily an easy matter. Cetaceans have evolved very streamlined bodies for efficient swimming. In the process they have eliminated anything that might get in the way—hair, hind limbs, external ears. They also couldn't afford to have their genitalia flopping about on the outside; hence everything is neatly tucked away when not in use, hidden in slits in their abdomens. If you get only a top view or a side view, or even too quick or too distant an underneath view, the animal's gender may not be apparent for most cetaceans. Some species show size differences between the sexes, but in most cases the difference isn't great; sperm whales and killer whales are notable exceptions. Male killer whales are easily recognized by their towering dorsal fins, which can reach heights of nearly six feet. *Tursiops* males run somewhat larger than the females, but given the range of individual and regional variation, size isn't a good indicator of gender among bottlenose dolphins.

The principal way for a human observer to tell whether a dolphin is male or female is by the number and arrangement of slits on the underside. In males the penis is concealed within a slit located roughly midway between the umbilicus and the flukes. (A dolphin's belly button looks like a big dimple or pockmark.) The penis is curved into an S-shape, tapering almost to a point at the end. Its erection is a voluntary act: male dolphins can pull it in or protrude it from the genital slit at will. No embarrassingly inopportune adolescent hard-ons, no unbidden nocturnal emissions—though I doubt dolphins would suffer much embarrassment over such matters, anyway. In fact, males will often use their erect penis as an appendage, carrying things around with it, even hooking it around a trainer's arm.

A male's anal slit is farther down the belly toward the tail, separated by three inches or so from the genital opening. In females the vagina and the anus are contained within a single slit. (At first glance Misha looked a bit like a girl in that regard, with only about an inch between his genital and anal openings.) Females also have two smaller mammary slits, one on either side of the genital slit, which house the nipples. (see diagram, next page)

Individual differences in appearance may be of greatest interest to field researchers such as Randy—that's what allows them to identify a large number of dolphins and follow their life cycles and social patterns. People who work with dolphins in captivity, however, are likely to be more interested in differences in character, in temperament. The real joy of it all is the opportunity to get to know individual dolphins close-up and one-on-one, to be a part of their lives day to day, to discover their moods, their quirks, their unique, complex personalities. Make that "dolphinalities."

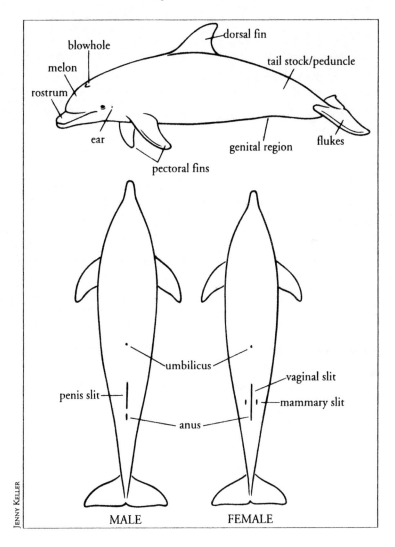

Echo and Misha were still impressionable youths when they arrived at Long Lab, and their "dolphinalities" seemed to develop, to take shape over the course of their stay. During their months in the thirty-foot pool, we began to notice a change in Echo and Misha's relationship, a shift in the balance of power between them.

Echo seemed to come into his own, to grow in confidence. Our daily training logs increasingly cited him as the pushier, more dominant one during feeding sessions. One time he even snatched a fish right out of Misha's mouth. Echo was the baby no longer.

Shortly after the first Lags arrived, we brought several additional people onto the crew (again, all female—whatever the reasons, that seemed to be the trend). The Boys already suffered a smaller tank; we didn't want them suffering from less attention as well, since the crew now had to divide its time between the Tursiops and the Lags. The new dolphinettes' initial assignment was largely to play with Echo and Misha between training sessions. They would dangle their feet in the water, toss a Frisbee or a football, run around the edge of the tank, talk to the dolphins, swim with them—anything they could think of that might engage or entertain the Boys.

The rest of us continued to play and swim with Echo and Misha as well. Swimming with them in the thirty-foot pool, though, we couldn't help but encroach on their space more than we had in the large tank, forcing them into closer proximity with us than they might have chosen. Misha often seemed put off by it. Echo started to get into it. He interacted with us more often and more extensively than Misha. He was more likely to toss the toys around or play a game of catch or tug-of-war with us. After a while he even came over to be rubbed. By late December he started giving us dorsal pulls, letting us hold on to his dorsal fin, and then speeding around the tank, person in tow. It was great fun from our standpoint and clearly seemed fun for him, too—at least he kept coming back for more. My best Christmas present that year was an exhilarating swim, full of dorsal pulls from both of them and much rub time.

Misha seemed more apprehensive about our presence in the tank. He was more likely to start chuffing when we were in the water with them. Occasionally he got somewhat aggressive, batting at a person's hands or feet, in a gesture that didn't seem entirely playful. He approached with an open mouth, in apparent threat, a couple of times. Some days he was more receptive, playing with the toys and coming in for rubs. He gave dorsal pulls, too, but less often and usually shorter than Echo's, with less apparent enthusiasm.

We introduced new activities into their training sessions to try to make their lives more interesting, things we thought would be fun, not just work, for them, such as doing leaps on cue and splashing water back at us. We also had a signal for "innovative," a "dolphin's choice" task: the dolphin would be reinforced for doing anything he wanted, so long as it was new. Ideally, it would be something he'd never done before, at least that we'd seen, but we accepted behaviors that were simply new to that session. Each time we gave the signal, the dolphin was to do something, any behavior, he hadn't done earlier.

Animal trainer Karen Pryor first devised the "innovative" task with two rough-toothed dolphins, yet another species. The task is unusual in that it doesn't involve training a specific behavior; rather, the animal must learn the concept "do something new." This concept formation is called second-order learning, or deuterolearning. Certainly it takes intelligence to form that sort of concept. But apparently pigeons have done it, too, so we shouldn't get too worked up about the sort of IQ it may reflect. Her two dolphins came up with some spectacularly original behaviors, elaborate leaps, corkscrews, back flips. Furthermore, they seemed genuinely excited, turned on by the concept, offering behaviors no trainer would ever have dreamed of.

Our two Tursiops never came up with anything quite so spectacular. They didn't have room enough in the thirty-foot pool for any fancy leaping, anyway. They concentrated more on performing a variety of actions on or with the various toys in the tank. Echo was much more imaginative about it than Misha. He also seemed to enjoy it, to be energized by it. He was especially good at coming up with subtle variations on a theme, testing the limits—just how different does the behavior have to be to be considered new? "If I swim around balancing a Frisbee on my head and carrying a hula hoop on my left pec, does it count as new if I switch the hoop to the right pec the next time around? How about if I change Frisbees?" He seemed to compete actively with Misha, even swiping toys from his tankmate to use in his own show.

Misha usually just repeated a few of their standard trained behaviors—fetching various toys or tossing them about, perhaps doing a leap—without any embellishment. He never really seemed to grasp the concept of "do something new." Or we failed to communicate it to him adequately. Or he simply didn't care for the game.

In general Echo caught on more quickly to new training tasks than did Misha. He was also more erratic in his willingness to perform them, however. Maybe he'd do it, maybe he wouldn't. Misha tended to be slower to get the idea but more reliable once he had it down.

Increasingly, Echo seemed the more outgoing, the more flamboyant of the two. Misha was quieter, more thoughtful. He worried more. He reminded Holly of those people who can't understand math because they overcomplicate it. "He'd never believe there are negative numbers," she concluded.

Misha was also the one who got zits. Bottlenose dolphins no longer have any body hair, but they do have a few remaining hair follicles along the rostrum. From time to time one of Misha's follicles would become infected, looking for all the world like a big pimple on his face. That's essentially what it was, and that's essentially how we treated it—squeezing out the pus and applying an antiseptic.

Echo was more of a brat. The two of them took turns standing "guard duty" during their nighttime rest periods. Misha was dutiful, diligent on his shift. Echo, when it came his turn, was a restless guard—you could almost see him checking his watch every few minutes. He made sure his companion didn't rest too soundly, either, bumping into him or nudging him awake periodically.

Echo liked to tease his trainers as well as his tankmate. Both dolphins soon learned they could get a reaction from people by getting us wet. They could breach—leap into the air and land full on their sides near the intended victim, sending out a great, drenching wave of water. Or they could just bring their flukes down squarely on the water in front of the trainer's face as she lay on the training platform. Most effective at close range. They seemed to know what they were doing, and to enjoy the effect. Especially Echo. He also took to nipping at our fingers—just a quick little snap that seemed designed more to evoke a response than to injure. He never caused any serious damage, though a bite occasionally broke skin. Misha didn't demonstrate the same experimental interest in human reactions Echo did. If he nipped, he was likely to mean it.

Ultimately we tended to think of Echo as the dominant one of the pair, though I'm not convinced it was so much a matter of dominance as of differences in temperament. Misha wasn't submissive; he didn't kowtow to Echo. They each stole fish from the other at times, and they both instigated bouts of rowdiness. They seemed pretty well matched in fights and tended to come away with about equal numbers of cuts and bruises.

The two of them seemed to get along amazingly well, considering they were in each other's faces all the time. They engaged in a lot of rowdy, rough-and-tumble play—like young boys wrestling, in sport and yet in earnest. Only once did I see what I'd consider a serious fight. During a training session in the thirty-foot pool, Echo apparently butted into Misha's way just one too many times. Misha reared up high and came down, mouth open, on top of Echo's head, giving him a good gash on the melon. The two of them tore around the pool, chasing and raking each other with their teeth, sending up a great spray of water. There was nothing we could do but watch, and hope they didn't hurt one another seriously. The skirmish ended in less than five minutes, leaving both of them with only minor injuries. That encounter somehow seemed angrier than the rest, without the usual element of playfulness. I'm not sure just why Misha lost his temper that day and not on so many similar occasions, but I guess everyone has his limits. In general, I was impressed by Misha's tolerance of his tankmate's antics.

Looking back, I can't help but wonder how much and in what ways captivity might have affected the development of their "dolphinalities." At this point I'm inclined to describe their individual temperaments in such terms as: Echo the Quick and Brash, Misha the Slow and Tentative; Echo the Impish, Misha the Equable; Echo the Mercurial, Misha the Temperate. All the crew members I talked to in preparing to write this book used similar terms, referring to Echo as the more dominant, the quicker, the smarter one. So strong was this sense of who they each were that we'd all forgotten it had ever been any different. Only Jill, so connected to Echo from the outset, recalled his early timorousness. It wasn't until I reread the early training records that I remembered how Misha had once seemed bolder, more sure of himself, how initially he had forced Echo out of the way during feeds and had generally been the first to swim through the passageway during gating sessions.

I don't think either of them actually underwent a major "dolphinality" change. Echo's impishness was evident from the outset—that's part of what Jill saw and responded to in him. I was drawn to Misha's brand of quiet strength, even from the days back in Ruskin. But something did shift over the course of their stay with us, and I suspect that shift, in part, reflected differing responses to their new circumstances.

I'm inclined to agree with Holly's assessment that "Echo understood the human condition better than Misha." I think he not only understood us better but also was better able to blow us off—or maybe those two go together. If he didn't understand what we were asking of him, or if he didn't feel like doing it just then, thank you, he'd say the hell with it. Misha struggled, trying to figure out what we wanted, trying to make up his mind what to do. He seemed more baffled by us and our demands than Echo did, or at least more concerned with trying to figure it out. For example, even though Echo initially appeared to be more terrified of gating than Misha did, Misha was perhaps more upset by the whole process, the going back and forth, over and over again—what *did* those humans want, anyway? I suspect he was probably braver than Echo—though not as foolhardy—when it came to the sort of life threats they were accustomed to in the wild. But he was perhaps less prepared to deal with the seeming craziness and unpredictability of people. Echo had enough craziness, enough unpredictability, of his own, I believe, to cope better with the strangeness of life with humans.

My sense of Echo's and Misha's "dolphinalities" no doubt derives in part from comparisons with Jo and Arrow. Josephine and Arrow were the first dolphins I ever really met; I'd seen some in zoos before, but that was through glass, from a distance. Jo was the first dolphin I ever touched, and something about her certainly touched me. She was my "first love," I guess, and still owns a good piece of my heart. She's got "dolphinality" to spare.

Josephine is a cantankerous old cuss. Arrow was perhaps more what I expected in a dolphin—sweet-tempered, sleek, and beautiful. Jo is lumpier, in body and in spirit.

Karen Pryor, on a visit to Long Lab as a training consultant, described bottlenose dolphins as the "juvenile delinquents" of the whale clan, and Josephine in particular as "a dolphin carrying a switchblade." The latter may be unfair. Jo measures nine feet long, weighs more than five hundred pounds, and believes in fighting back; but she isn't vicious. It's just that anyone with a "flower child" image of dolphins had best look elsewhere.

Jo is a survivor, tough and streetwise. She has the look of a dolphin who's been around, something of that "hooker with a heart of gold" air. When I knew her, at least, Jo had a pronounced double chin—she does love a good mackerel. She has a notably bulbous melon. Her dorsal fin cants to the left. A row of jagged scars mark where some long-ago shark once planted its teeth in her back. She has a baseball-sized scar, white and nearly round, of unknown origin, along her left flank. Plenty of smaller scars and nicks further attest to her years and her experience. The lower tip of her rostrum is white, a common trait among older dolphins, the pigment apparently worn away from rooting around on the bottom, sticking her snout into things.

Jo is also exquisitely expressive. She made a distinctive little chuckle—a self-satisfied "nyeh, nyeh, nyeh"—whenever she got a mackerel. But if things weren't going her way during a training session, she'd let fly with what I can describe only as a string of expletives befitting a twenty-year Navy veteran.

She was also master of the tailslap. Done forcefully, it can sound like a rifle shot. Sometimes she'd give us a modified version, turning and showing her tail in a gesture that, if a dolphin had fingers . . . Well, we translated it as "fluke you."

I had never before reacted to any other animal quite the way I reacted to Josephine—and I've been drawn to animals of various sorts all my life. I never before felt so strong a desire to understand, to know what was going on inside her head, to get at least some glimpse of the world as she saw it.

Josephine is a thinker. She seemed to get interested, to work at figuring out what it was we were asking her to do. And it wasn't just a matter of getting fed for doing so. When she didn't feel like cooperating, no amount of fish could coax her. At other times she would work but wouldn't take the fish we offered. Jo is her own dolphin—she can't be bought.

When we did routine physical exams, we moved the animals into a smaller holding pool and then drained the water. Arrow generally just closed her eyes throughout the procedure and seemed almost to fall into a trance. You'd swear she was chanting

"Ommm." Not Josephine. Jo is a fighter. She never really hurt anyone—and I don't think she ever really tried—but even with a dozen people holding her, she could still give us a good ride if she started swinging her flukes. More than that, though, she simply *had* to watch. When we were working on Arrow, Jo would maneuver her body around on the tank bottom so that she could see what was happening; then she would lie quietly.

In some ways Jo and Arrow presented a contrast similar to the differences between Echo and Misha. During physicals Echo was more likely to fight, as Jo was; Misha, like Arrow, was more apt to close his eyes and pray for the whole business to end soon.

Jo, like Echo, is a more flamboyant, outgoing, take-charge sort of dolphin. She lets you know how she feels about something, clearly. She is highly vocal, with a wondrous array of sounds for all occasions. Arrow was quieter, gentler. If you had to describe the two of them with respect to dominance, certainly Jo would be the dominant one, though again I have my doubts about the term. Arrow wasn't so much submissive as withdrawn. Much of the time she seemed almost to live in a world of her own. Whether that was her "dolphinality" or perhaps a reaction to captivity, I'm not sure. Or maybe it was simply a matter of ill health; it's hard to be fully present when you're not feeling well.

Arrow generally seemed less responsive, less interested in people than Jo was. Jo showed more interest in us, but more mistrust as well. I suspect she'd been through a lot in her twenty years of captivity; she certainly had been moved around a number of times, changing not only her surroundings but the people and the other dolphins in her life as well. Maybe she just never knew whom or what she could count on.

Echo and Misha seemed like more of a real pair than Jo and Arrow in the sense that they were truly in this thing together; their experience had been a shared one from the outset, and, for good or bad, they were bonded by it. Jo and Arrow, on the other hand, had been captured separately, at different times and places from each other. They had been through largely different experiences in captivity and had been with a variety of other animals before being paired up at LML.

Jo and Arrow did appear to like each other and to get along well. They were never as rowdy with each other as the Boys were. They were sexual with each other but never aggressive. Jo even seemed rather solicitous toward Arrow. One time after a feed, Arrow stayed at the platform and showed far more interest in being rubbed than she usually did. Jo didn't seem to approve. She swam by and chased Arrow away from the platform. Arrow kept returning for more rubs and Jo kept chasing her away. Michelle and I were both on the platform, and we offered rubs to Jo as well. She wasn't interested. She just seemed intent on preventing us from rubbing Arrow. Finally, since both the humans and Arrow seemed to be ignoring her demands, Jo came over and whacked my hand with her rostrum. Not hard enough to hurt me but enough to get my attention, to let me know I'd been hit. We desisted, honoring Jo's wishes, even if we didn't entirely understand what they were. I don't think she was jealous—she could have gotten all the rubs she wanted. It seemed more as though she were trying to protect Arrow, warning her not to get too cozy with those quirky humans.

The Lags add yet another point of comparison, this time contributing differences in species and ecological niche to the "dolphinality" picture. Lags can be differentiated physically on the same general bases as Tursiops—dorsal fin, size, coloration patterns, scars. Eleven, the male, was simply the biggest of the four Shedd Aquarium Lags, with a large, very hooked dorsal fin. Ten and Twelve were considerably smaller. Twelve had a very pretty melon, like an abstract black-and-white airbrushed painting, and a large white splotch on her right flank. Ten had very pronounced "suspenders" and a mottled scar on the side of her mouth. Nine had "eyebrows,"

white patches above each eye. She was also something of a loner, more apt to swim apart from the group than the others, and given to peering at herself at great length in the underwater viewing windows.

It's easy enough to differentiate between Tursiops and Lags in terms of their shape and coloration patterns—no one ever asked, "How do you tell the two species apart?" But there are more subtle differences as well. They move rather differently, for one thing. The Lags are like bullets with tails; except for the up-down beat of the flukes, their bodies essentially move as a unit. The Boys are more flexible, wonderfully fluid in their movements, their bodies arching and curving up, down, and sideways, in all dimensions at once.

The two species feel different, too. Visitors to the lab often asked what dolphins are like to touch. We all had our own answers. Michelle, true to her Mediterranean heritage, said they feel like giant olives. Others cited wet leather or an inner tube. Holly suggested the inside of a person's forearm. The Lags felt harder, more taut than the Tursiops. In Holly's description the Lags were a more muscular forearm, the Boys a bit flabbier. (I prefer to think of them as supple, not flabby.) If you opt for the inner-tube image, the Lags had more air in them. Or, in terms of olives, Lags have pits, Tursiops pimientos. None of the analogies really does any of them justice, though. All the dolphins were silky smooth and luxuriant to touch—except when they were sick. Illness seemed to deflate them, turning those strong, firm bodies temporarily soft and flaccid.

The Lags were certainly much noisier than the Boys. They were squawking and screaming much of the time. The Boys, except during their jousting bouts and during training, tended to be fairly quiet. That could be simply a matter of numbers—there's more to say when there are four of you than when there are only two. In places where they are more numerous (such as the DRC in Florida), Tursiops do seem more vocal than Echo and Misha were. Individual differences may also play a part: Echo seemed more vocal than Misha, Jo noisier than either of them.

There is probably a species difference as well. Lags tend to live in large schools; they are always dealing with many other individuals, in groups that are constantly changing in composition. They need to keep the channels of communication open to keep track of everybody and to establish and maintain their relationships properly. Bottlenose dolphins—coastal ones, anyway—tend to live in smaller groups, and they may not need to gab as often with as many different animals as Lags do.

Echo and Misha seemed, in general, more cautious, more conservative than the Lags; the Lags were more adventuresome, more willing to try something new. When we put something unfamiliar in the tank with them—a new toy, recording equipment, or some training device—the Boys tended to shy away from it, whereas the Lags fairly quickly approached the novelty, soon started playing with it, and often managed to demolish it. One time we tried a large plastic frog, a wind-up toy that would swim through the water. The two Tursiops fled, swimming to the opposite side of the pool and facing the wall. The Lags, after a brief hesitation, approached the frog, scrutinized it, swam around it, echolocated on it, even touched it. They seemed more fascinated than frightened.

Again, the difference may partly be a matter of numbers—it's easier to be bold when there are four of you than when there are only two. It may also have to do with species differences in their sociality. Coastal bottlenose dolphins, who are often in small groups and may even be on their own at times, have to think of self-preservation in individual terms—they just might find themselves one-on-one with a shark. For Lags, on the other hand, the group serves as their protection; it provides their warning network and their camouflage. They may not need to exercise their caution on an individual basis quite as much as Tursiops do.

Maybe Echo and Misha were more apt to consider everything they encountered as a potential threat because their home environment was potentially more threatening. The Lags lived in open ocean, where predators had nothing to hide behind, and where there wasn't much else to harm them. Tampa Bay, on the other hand, not only provided ample hiding places for sharks, it also offered no end of other things a dolphin could worry about: obstacles in the water, shallows, the tide schedule—if you go in too shallow when the tide is going out, you could get stranded—plus large numbers of people, with all their motorboats and their potentially hazardous refuse. The Boys may have learned caution at an early age.

The Lags seemed more tolerant than the Tursiops, not only of new things in the tank but also of changes in routine and just plain old accidents. The Boys tended to get mad, chuffing and tailslapping, when we were late with a feed. They also got mad if we changed their food in any way. The Lags, too, would get upset—when we cut their fish, for example—but without quite the anger, the sense of moral outrage, Echo and Misha managed to convey.

People fell into the tank accidentally while vacuuming or standing on gates. A couple of times one of the training platforms broke during a session; the dolphins found this understandably upsetting, as the platform plunged into the water in front of them, spilling people and buckets as it went. Later on we sometimes deliberately pushed people into the tank, our ritual farewell when someone was leaving the crew. We would throw someone in only with the Lags, however, not with the Boys. The Lags seemed to take it all in stride—they would dash off as the person splashed into the water and maybe swim fast around the pool a couple times. Then they were fine, even curious. Echo and Misha were more likely to sulk, again swimming to the opposite side of the pool and facing away. Sometimes they even refused to cooperate during a subsequent training session.

Echo and Misha seemed to hold us responsible, rightfully so, I guess, even if whatever happened was accidental. The Lags seemed to have a more accepting, "shit happens" attitude. They would simply shrug (if an animal without shoulders can be said to do that) and go on their way without giving the matter further thought. The

Boys thought about it. They could hold a grudge for hours, sometimes days.

I never had the sense the Lags were genuinely angry at us, even
when they might be baffled or frustrated during a training session.
The Boys, however, clearly seemed to get mad, both at individuals—say, at the person working with them during a session—and
more generally, en masse, at humans, at the keepers of the tank.
Holly, whose interests lay more with dogs than with dolphins, said
the Tursiops reminded her of pit bulls, the Lags of golden retrievers.
She always felt more comfortable with the Lags, she said.

The Lags seemed to treat us almost like members of their
group, to accept us as part of their social structure. Echo and
Misha, while certainly willing to interact with us (more so at some
times than at others), seemed to regard us more as aliens.

That difference may stem, in part, from their differing social
ecologies. Lags in Monterey Bay frequently associate with several
other species of cetaceans, often forming mixed-species schools.
Their most frequent companions include Northern right whale dolphins, slender, fast-swimming creatures, with no dorsal fin and tiny,
graceful flukes; Risso's dolphins, light gray colored, somewhat
larger than a Tursiops, with a blunt head and no noticeable beak;
and Dall's porpoises, enthusiastic bow-riders, with dramatic black-
and-white markings and stocky little bodies. Lags are accustomed to
accepting other species as part of the group, to interacting with
them as schoolmates.

Tursiops in some areas do associate with other cetacean species.
Echo and Misha, however, were unlikely ever to have seen any other
cetaceans, much less to have hung out with them on a regular basis.
Spotted dolphins are found in the Gulf waters, not far away, but
the only other marine mammal in Tampa Bay is the Florida manatee, a slow-moving vegetarian who has little in common with the
dolphins. For Echo and Misha other species had always been
predators, prey, or playthings, never friends or colleagues. Dealing

with us was a whole new class of interactions for them. Like humans, perhaps, they had difficulty viewing other species as equals.

Lags are true schooling animals, highly grouped, like a herd of buffalo or a flock of starlings. Their "dolphinalities" are intricately interwoven into that of the larger group. Dawn describes the Tursiops as whole individuals, the Lags as parts of a larger organism—the school. Bottlenose dolphins seem to function more independently, as individual decision makers. Like us, they are sociable but relatively autonomous.

The Lags seemed more spontaneous, the Tursiops more intellectual; they mull things over. The Lags take a more trial-and-error, go-for-it approach to life. "Lags are like the stuntmen of the world," Holly said, "Tursiops more like theologians or monks."

"No, they're too horny," she amended, "make that twisted monks."

Karen Pryor is not alone in regarding bottlenose dolphins as juvenile delinquents, perhaps even with a malicious streak. Wild Tursiops have been known to drown sea turtles, apparently for sport. That still may be their intellectual side: "I wonder what will happen if I hold it down . . . ooh, it wiggles a lot, I wonder why . . . huh, it's stopped wiggling . . . boring."

An older male Tursiops Michelle had previously worked with apparently experimented on the pipefish—a thin, tubular relative of the seahorse—that regularly swam into his bay pen. He would take some unlucky fish into his mouth and try to plug it into all manner of holes or crevices around the pen, seeing what it fit into. Intellectual curiosity.

Sex seems to be a common denominator among all dolphins, both captive and wild. They don't do it just in the interest of procreation, either. While they do breed seasonally, generally mating in spring or fall, and sexual activity may reach a peak then, it is by no means confined to these times. As Ken Norris says, "Dolphins use sex like we use a handshake." It's the dolphin national pastime. Their sexual behavior extends to members of both genders and all ages, to inanimate objects and, at least under some circumstances, to members of other species. It's not necessarily sex in quite the way we think about it, however. Sexual interactions are an integral part of dolphin sociality, part of how they affirm and reaffirm their relationships.

Bottlenose dolphins, highly adaptable in so many regards, have a way of blurring even the lines between species. "Species" refers basically to a group of individuals that are capable of interbreeding. In captivity, at least, Tursiops have been known to breed successfully with various other delphinids, including rough-toothed dolphins, pilot whales, false killer whales, and Risso's dolphins. What's more, some of these matings produce fertile offspring, unlike the horse-donkey cross, which results in a sterile mule. A "wholphin"—born of a Tursiops mother and a false killer whale (pseudorca) father at Sea Life Park in Hawaii—later gave birth to a calf of her own, fathered by another bottlenose.

I've heard reports of humans in some of the swim-with-dolphins programs being accosted by libidinous dolphins. During their breeding season, male Tursiops may give an especially hard time to human females who are menstruating. We tried to make it clear to Echo and Misha that such behavior was unacceptable. Sometimes when they came by to be rubbed, they would offer up their genital area; we refused, waiting until they presented some more suitable part for rubbing. We wouldn't venture into the water with them at all if they were in a particularly randy mode.

The Girls, the Boys, and the Lags all engaged in a considerable amount of sexual behavior. They spent a great deal of time gently caressing each other. They could also play rough. Echo and Misha engaged in regular jousting matches, ramming into each other head-on, whacking each other with their flukes, and tussling about, mouths open. They would rake each other with their teeth, sometimes leaving a row of gashes in their partner's skin. That was foreplay.

The Lags were perhaps even more active sexually, since there were more of them, and they had both genders covered—more possibilities. They, too, played rough. The male, Eleven, chased after all the females, but he was especially persistent with Nine. At times her body was nearly covered with rake marks. One time Eleven kept chasing her mercilessly, and she kept leaping and leaping—the only place she could really get away from him was in the air. Finally she leapt clear out of the tank and skidded across the deck. She was unhurt, if terrified. It scared the bejeebers out of the crew as well. Fortunately, it happened during the day when there were people around to get her back into the tank. We had taken down the netting around the tank because the Lags had been doing so well. After Nine's big leap we brought it out again, putting the netting up around the tank each evening before we left the lab and taking it down again the following morning.

Dolphins, evidently, are capable of what we would consider sexual harassment. That goes for wild dolphins as well as captive ones. Researchers who have been studying the bottlenose dolphins in Shark Bay, Australia, report that the males there form coalitions, similar to the pair bonds and trios Randy has observed among Sarasota males. One function of such coalitions apparently is to sequester females in estrus: working as a group, the males can force their attentions on a female and prevent her from escaping. This sort of behavior appears to be much less frequent among Randy's Sarasota animals. That disparity could simply be a result of differing observational techniques between the two studies. On the other hand, *Tursiops* do exhibit regional variations between groups in feeding behaviors, so perhaps there are "cultural" differences in sexual mores as well.

The Long Lab dolphins all played with a variety of toys as well —assorted balls, Frisbees, plastic baseball bats, toy boats, boat buoys, hula hoops—anything we could think of that might be interesting to them and which they couldn't hurt themselves on or demolish immediately. At least some of the play was sexual in nature. Jo and Arrow especially liked the bats, which we'd partially filled with water so they were submersible; they treated the bats something like giant dildos, pushing them to the bottom of the tank and rubbing on them. (We always had to take the toys out of the tank before a training session, or the dolphins might be distracted.) Arrow also liked swimming around with a Frisbee against her genital area, somehow held in place by the force of her movement. Some of the Lags discovered the same trick. Echo had a favorite plastic buoy we started calling Buddy because he spent so much time with it—mostly carrying it around with his penis.

Another big hit—and one the Lags, at least, may have seen before—was kelp. We'd walk down to the beach, pick out a couple of long strands of kelp, and toss them into the pool. The kelp wasn't exactly a sex toy but seemed to have some aesthetic appeal. This particular variety looks like a long feather boa, and that's how the dolphins used it. They would swim around with a kelp boa draped over one pec, or wrapped about their flukes and trailing out behind them. Arrow especially loved to swim with the kelp, as did a couple of the Lags, primping and prancing around the tank. You could almost hear the strains of "I feel pretty . . ."

Dolphins are highly social and highly sexual. They are curious and playful, and I suspect most dolphin trainers would swear they have a sense of humor. They have individual temperaments; they have their moods, their likes and dislikes. They have their friends; they have their spats. You have to love them for their sassiness as well as for their sweetness, as much for their irascibility as for their affability. Lilly's "humans of the sea" notion isn't altogether unfounded.

5

Making Common

Referring to dolphins as "humans of the sea" may help us appreciate some of the similarities between ourselves and these marine mammals. But the phrase also ignores just how different dolphins are from us, and how different the sea is from the world in which we live. The enormous disparity between these two worlds seriously restricts the extent of communication possible between our two species, no matter how smart dolphins (or we) may be. The word "communicate," after all, derives from the Latin *communicare*, meaning "to make common." One of the advantages of working with dolphins in captivity is that it establishes a basis for communication by creating a shared world, one common to both humans and dolphins.

I continue to have some qualms about seeing dolphins (and other animals) in captivity, and I wholeheartedly believe we should learn as much as we can about them in the wild. We are getting better at finding ways to be a part of their world. Still, the more I've worked with dolphins in our world, the more I've come to value what we can gain through this form of "making common" as well.

For one thing, some matters simply aren't amenable to field research. We would know little or nothing about dolphin echolocation, for example, without laboratory studies. More important, I think, working with a captive dolphin can be something like a marriage. Granted, from Echo and Misha's standpoint, it was an arranged marriage. But the daily interactions, the regular physical contact, the understandings and misunderstandings, the sharing of both affection and frustration, can give you glimpses into the heart and soul of another being—and into your own as well—that simply can't be gleaned from more distant observations.

One of our principal modes of communication with dolphins in captivity is the training process itself. Many people tend to think of training as simply a matter of teaching the animals tricks, or of making them conform to our wishes. Not so. I don't think that's true in any kind of animal training, at least not when you're using positive-reinforcement training, but there is certainly no way you can make a five-hundred-pound dolphin in the water do something it doesn't want to do.

Animal trainer Karen Pryor refers specifically to "reinforcement training as interspecies communication." Training of this sort functions as a two-way system, an interactive process in which what the trainer does depends on what the animal does. She points out that "as the trainer is developing behavior, the animal in effect is training the trainer to give fish."

The Boys played their own games of "train the trainer" with us. One favorite routine, especially for Echo, was to get the trainer to lean out farther and farther over the water. When I had him station with his rostrum resting on my hand, I initially placed my hand in the water very close to the training platform. But then gradually, subtly, he'd drift back away from the platform, until I was practically falling into the tank to keep my hand in place under his snout.

He suckered me into that one repeatedly. I wasn't the only one to fall for it, either. Michelle regularly had to point out to one or another of us, "Look where he's got you."

That trick always reminded me of an apocryphal story from my undergraduate days in psychology about a class that tried to train their professor to jump out the window. The students all conspired to pay close attention to the lectures only if the teacher stood on the side of the classroom nearest the windows; if he strayed into the other half of the room, their attentiveness lapsed. The professor, responding to the students' apparent enthusiasm, spent more and more time on the window side. Gradually the students moved their criterion point closer and closer to the windows. They had the poor guy lecturing from the windowsill before he realized something was up.

With Echo and Misha we worked on training three general categories of behaviors. First, we needed them to perform certain tasks specifically for the echolocation research. One of the most important, and most difficult, of these was convincing them to permit us to blindfold them: to ensure that they would indeed echolocate, we placed a soft latex suction cup over each eye.

Second, we worked on various husbandry procedures that would allow us to monitor their health. From the outset we wanted to do daily eye inspections, mouth inspections, and overall body checks. For this they needed to bring their heads up out of the water and look us in the eye, one eye at a time, or open their mouths so we could look and feel inside, checking their tongues, teeth, and palates. They also learned "parallel stationing," in which they would swim by us slowly (when they were cooperating) and roll over so we could look at their whole bodies, checking for new rake marks, sores, or abrasions of any sort.

Later on we also introduced some more hard-core husbandry

behaviors. Specifically, we wanted them to present their flukes to us and hold still so we could draw blood; to open their mouths and allow us to put a tube down their throats and into their stomachs to obtain a gastric sample; and to give us a forceful exhalation into a petri dish so that we could get a blowhole sample. Dolphins at some facilities have been trained to accept these and other procedures (such as taking urine and fecal samples) quite readily. I've seen Marine World dolphins actively vie with each other for the opportunity to get the stomach tube—a matter of attention, I guess. Again, dolphins have no gag reflex, and they routinely swallow whole fish much larger in diameter than the tube, so if the animal is relaxed and willing, the process is no big deal.

In principle all this isn't much different from our learning to sit still in the doctor's office and allow ourselves to be poked, prodded, and the like. The dolphins may not know we're doing it for their health, but they can generally learn to accept it willingly enough. It's certainly much easier on the animals, as well as on the people, if the dolphins will voluntarily permit such procedures rather than having to be pulled out of the water and physically restrained. More and more facilities are incorporating such training into their routines, not only with dolphins but with other animals as well.

The third general training category was "fun stuff," behaviors that by and large were easy for the dolphins to pick up and that we figured would be enjoyable, both for them and for us—something to give us all a break from the rigors of the research and husbandry training. This involved things like the "innovative" sessions described earlier, leaps, fetch, pec waves, "yes" and "no" head movements, and a lot more—many of them developed from behaviors the Boys themselves offered first.

We started training "yes" and "no," for example, in response to a head-bobbing action Echo did spontaneously. We shaped the behavior into a more clear-cut nod, later adding the head shake, and had a good time asking the Boys questions to which they could answer yes or no. Echo tended toward a rather manic technique, nodding his head up and down several times very quickly—a sort of "yes, yes, oh yes" response, whereas Misha gave a slow and easy "yup."

We kept detailed logs, one for Echo and one for Misha, of all our training-feeding sessions. We recorded the date and time, who worked with which animal and where, what they worked on, how the dolphin performed, the kind and amount of fish he ate, and anything unusual or interesting we might have observed.

The first step in training is to get the dolphins accustomed to the whistle. Before you can train them to do anything, they must learn to associate the sound of the whistle with something good—a fish, being rubbed, toys, whatever they happen to like. Generally, you primarily use fish at the outset, especially with a newly caught animal that may still be skittish about being touched by humans or about strange objects in the tank. With Echo and Misha we'd started pairing the whistle with fish from day one, while they were still in the pen back in Ruskin.

The whistle is called a bridge, because it's used to bridge the time delay between whatever behavior the dolphin does that you want to reinforce and the actual, tangible reinforcement (fish, etc.). If, for example, the dolphin spontaneously does a nice leap, which you'd like to train him to do on cue, by the time you manage to give him a fish for it, he may have done ten other things in between—blown a bubble, taken a breath, swum across the tank, defecated—and may have no idea which of them you're rewarding, or may simply associate the fish with whatever he did last. (Okay, it can be useful for husbandry purposes to have the dolphin trained to defecate on cue, but that's probably not where you want to start.) The whistle allows you to give a signal immediately, as soon as the animal performs the behavior. And since the sound of the whistle has been associated with fish and other goodies, it has become reinforcing in and of itself (called a secondary reinforcer). It bridges the time gap between the desired behavior and presentation of fish, or anything that is rewarding to the animal without prior training (a primary reinforcer). So the animal learns that whatever it was doing when the whistle sounded was something desirable and will lead to goodies. Sometimes it may take many associations of behavior-whistle-fish before the dolphin makes the connection, sometimes only a few or even just one, depending on the behavior, the dolphin, and the trainer.

A good trainer needs to be very precise with the whistle. You've got to be quick enough to make sure you're reinforcing the behavior you want, and preferably catch it at its height, in its best form. If you're trying to get a high, clean leap, for example, you want to blow the whistle when the dolphin is at the highest point of his leap, not when he's coming back down into the water. Also, once he understands the general idea, you want to reinforce only good leaps—ones that meet a certain criterion, or ones better than the last one. If, after he's been doing proper leaps, you then accept a sloppy one, you'll be reinforcing a more slipshod performance than you want. In the early stages of training you may accept less impressive behaviors than you ultimately hope for, however, and then gradually try to shape them into something more elegant.

Once the dolphin has the basic idea of the bridge, you can start trying to get certain behaviors on cue. You don't want to reinforce him whenever he decides to leap but rather when he does it on cue from you. You can also begin to shape new behaviors, things the animals don't necessarily do spontaneously, or at least not quite in the form you'd like it.

To help improve our training skills, the crew members sometimes played "the training game" among ourselves. Among other things, this gave us a glimpse of what the process feels like from the dolphin's standpoint. I was blown away the first time I played. It's far more difficult than you might imagine to figure out what in

God's name the trainer wants when all you have to go on is the sound of the whistle when you do something right.

For a given round, one person would act as trainer and another person would act as dolphin. Usually we had a bunch of toys and props in the room, as we might have in the tank, at least for "innovative" sessions. First the "dolphin" leaves the room briefly while the trainer and the rest of the crew decide what she should try to get the "dolphin" to do—build a tower with blocks, run a toy truck around the front of the room, stand on one leg and touch her nose with her left hand, or maybe even just stand still (which could be hardest of all to train) with her hands behind her back. Then the "dolphin" returns to the room, and the trainer tries to shape the desired behavior, using nothing but the sound of the whistle.

With a real dolphin some behaviors, such as a leap, can simply be "captured"—that is, the dolphin does the behavior naturally and all the trainer does is try to get it on cue and perhaps sharpen it up a bit. Then you can just wait for the dolphin to leap on his own, blow the whistle, and toss him a fish when he does so. Often the animal will get the idea with only a few repetitions of that sequence. Then you can start holding out for higher leaps or ones that cut the water cleanly. Later you even start working on a discrimination between leaps—relatively splash-free, nose-first reentry—and breaches —a belly flop or side flop in which the animal hits the water with greater body surface and sends up a great spray. Dolphins do both behaviors naturally; it's getting them to discriminate between the two and do the appropriate one in response to the appropriate cue that may take some work.

Other behaviors aren't so natural—many of the research-related tasks, for example, or the various husbandry procedures. You can't just wait around for a dolphin, of his own accord, to present his flukes for a blood sample. Then you need to shape the behavior more or less from scratch, using a process of successive approximations. You start off by reinforcing—blowing the whistle—a behavior that bears some resemblance, however remote, to the behavior you ultimately want to train. That may simply be a matter of facing in the right direction at the outset. Then gradually you hold out for something a little bit closer to what you want, working toward closer and closer approximations to the desired behavior. Generally, if the animal doesn't get reinforced for doing whatever it had just been reinforced for, he will start playing with the behavior, changing it a bit here or there, trying to bring back the reinforcement. The trainer then needs to pick and reinforce one of these alternatives that seems a bit closer to the goal than the last behavior.

In the training game, for example, the human "dolphin" may start by moving her arms or legs or head, or by walking around the room. The trainer then blows the whistle for the first movement of a particular body part or a step in the proper direction. Once when I was dolphin, the desired behavior was for me to pick up a toy frog off the table and place it inside a teacup. My first whistle came simply for turning in the direction of the table. Then for walking toward the table. Then for picking up objects on the table. I got confused when, after just having been reinforced for picking up a yo-yo, I wasn't reinforced when I picked up the yo-yo a second time. Repeating the reinforced behavior—or what you thought was being reinforced—is a way to double-check to see if you were right; if the behavior is reinforced again, then you know you're onto something. That's a common response among dolphins as well as people.

I had the right idea—picking up objects—but I hadn't yet performed the goal behavior. I had to keep trying. First I tried doing different things with the yo-yo, thinking maybe it was the right object but the wrong action. When that didn't seem to work, I went to other objects, picking them up, playing with them, testing all the possibilities I could think of. Finally I picked up the frog. The whistle blew. I started playing with it. When I brought it near the teacup, the whistle blew again. I placed the frog inside the cup.

Stronger whistle. I took the frog out and did it again—another whistle. Again—whistle. Finally we all knew I got it.

The kind of confusion I felt when I wasn't reinforced for the behavior I'd just been reinforced for a minute ago is very common in training. There must be times when the dolphin is terribly confused and frustrated: "But you liked that the last time I did it!" A good trainer needs to be able to tune in well to the dolphin's frustration levels and to be adept at breaking the desired behavior down into manageable steps. You can't fairly expect the dolphin to make great intuitive leaps between one step and the next, though sometimes they will do just that.

My confusion as dolphin that day was relatively mild. The situation was much worse during my turn as trainer that same day. I was trying to train the "dolphin" to approach the people seated in the first row, walk along in front of them and pat each one in turn on top of the head. Sounded simple enough. Getting him to pat one person on the head was relatively easy. Getting him to go from that person to the next and on down the line proved nigh onto impossible. He got hung up on trying different variations on the one pat, adding twists and turns to the arm movement rather than adding more people to the sequence. It took us a good hour or more, with some three or four time-out periods in which the "dolphin" was sent out of the room and the rest of us consulted about what I should try next.

I'm not an especially good trainer, I'll be the first to admit. Apparently, as the "dolphin" explained later, at some critical juncture I'd mistimed the whistle, inadvertently reinforcing the wrong thing and sending him off on the wrong track. Once he got into that mind set, he found it difficult to break, a common problem in training.

My "dolphin" was also a rather unusual one. He thought too much. He'd just stand there, thinking, for long periods of time, trying to figure out his next move, without taking any sort of action. That left me nothing to reinforce—you can't reinforce thoughts, only behavior. An occasional dolphin may behave that way—Misha did, to an extent—but most don't. By and large, if something doesn't work, they will try something else. Or dolphins will simply quit on you. The best "dolphins" to work with in the training game—and, I suspect, the easiest real-life dolphins to train —keep moving, keep trying things, presenting more possible behaviors for reinforcement.

This session was unlike what we'd be likely to do with actual dolphins in several regards. Given the lack of progress and the increasing frustration, a trainer in a real session would most likely drop that behavior for the time being and work on something else, perhaps even end the session. In general, though, you want to end a session on a positive note, not with everybody frustrated, so you'll probably try something the dolphin will succeed at before calling it quits. Then you can try the problem behavior again in a later session, or even the next day, after both trainer and dolphin have had a chance to relax. The dolphin may in the interim "unlearn" or let go of whatever erroneous response it had fixated on. Or you may come up with a better way to convey what you're after.

The sort of time-out procedure we used in the session is quite common in training. Sometimes you use time-out as we did in the game session, when whatever you're doing isn't working and the trainers want to go off and discuss strategy. More often you do it when the animal clearly isn't cooperating—won't come to station, breaks away repeatedly, harasses the other dolphin, and so on. Sometimes we'd just do a quick time-out, backing away from the water and turning our backs on the dolphins so as not to give any sort of reinforcement—such as attention—for the unwanted behaviors. Other times we walked off the deck altogether, staying away for five or ten minutes or so, sometimes longer. If you stay away too long, however, that becomes more like ending the session than a time-out.

Time-out isn't intended as punishment. All our training (and dolphin training generally, in most facilities these days) was based entirely on positive reinforcement. We didn't use punishment of any sort or withhold food from the animals. We might end a given session early, but Echo and Misha were always offered their full fish allotment during the course of the day. Time-out is simply a matter of *not* reinforcing some undesired behavior, or just giving the dolphin a chance to settle down a bit before proceeding.

Of course, two can play that game. Misha especially tended to time us out. If he wasn't happy about how the session was going or didn't feel in the mood for a particular task, he'd swim to the opposite side of the pool and turn his back on us, hanging there in the water facing the wall. Occasionally he'd glance back over his shoulder, apparently to see if we'd gone away yet. Josephine was also fond of giving us the "gray-back treatment," ignoring us if she didn't care for our behavior.

The training process may be somewhat limited in its scope, but it can provide an effective means of two-way communication between species. It's a set of rules or agreements: if you do this, I'll do that; when I do thus and so, I'd like you to do such and such. It's a way of creating a common vocabulary, a common set of understandings.

The trainer's vocabulary generally includes the whistle, of course. You can use several varieties of whistle patterns, such as our three toots for starting a session and a longer blast for a job well done. There are often obvious differences in the enthusiasm with which the trainer blows the whistle on different occasions as well. Sometimes the trainer may also employ an alternate sound signifying "no" or "you got it wrong that time," like the buzzer on a game show. Most dolphins don't seem to appreciate being told they made a mistake, however. Echo and Misha were no exception; they were more likely to swim off in a huff if we told them "no" than if we simply didn't reinforce them for an incorrect trial. By and large, we just went with "yes" and "no comment."

The trainer also uses a variety of hand and arm signals to indicate the particular behavior or sequence of behaviors desired. (Some facilities may use auditory signals instead, or in addition.) These signals are largely arbitrary and generally differ from one facility to the next; they aren't formally codified like the sign languages used by deaf people. Our signals generally involved larger, more sweeping movements than those in human sign language and were chosen for their visual distinctiveness—they have to be different enough from one another to not confuse the animal.

Some signals may resemble the action they indicate, a visual onomatopoeia of sorts. For example, we'd move a closed fist up and down at the wrist to signal the Boys to nod their heads yes, or wave a hand from side to side to ask them to wave one of their pectoral fins. Other signals, especially those for husbandry behaviors, often involved touching, tapping, or somehow manipulating the parts of the dolphins you want them to do something with. For example, we'd tap the end of Echo's or Misha's rostrum when we wanted him to open his mouth.

So the trainer uses a repertoire of hand and arm signals to let the dolphin know what she'd like him to do and uses the whistle to give him feedback about his performance, which also tells him whether a fish or other reward is forthcoming. For that matter, the number, size, and kind of fish given can convey information about the kind of job they did. We tended to give "jackpot" prizes, showering the Boys with fish or perhaps offering a favorite variety for a particularly good response.

The dolphin can, of course, indicate that he understands simply by doing the proper behavior. Sometimes, however, he may do the right thing without knowing it, so a correct response doesn't necessarily guarantee understanding. An incorrect response can be even more difficult to interpret. Maybe if you gave one signal and he did another behavior, he misunderstood. Then again, maybe he didn't feel like doing your behavior, preferring instead to give you one of his own choosing.

Among the "fun stuff" we worked on with the Boys was the tailwalk, a show-type behavior, seen often in old *Flipper* shows. In this maneuver the animal propels himself upright and backward through the water by fast, powerful beating of his flukes. For a while after he first learned the trick, Misha thought tailwalks were the be all and end all. Regardless of what we asked for, he'd respond with a tailwalk. When we gave the sign for innovative, he'd come back with a tailwalk—even if he'd just done one the last five times in a row. He seemed so pleased with himself, though, it was hard not to reinforce him for it.

Misha seemed, at least in this case, mostly to be showing off what he'd learned. Echo at times seemed more calculating. He was more apt to try to get out of something he didn't feel like doing by offering something else in its place. During one session when I was trying unsuccessfully to get him to do a slow, calm parallel station, he broke away, swam to midpool, and performed three quite lovely leaps in rapid succession. Then he dashed back, popped up in front of me with his mouth wide open, apparently anticipating a fish reward, with a look that seemed to say, "There, now, wasn't that better than some dumb old parallel station?"

Occasionally an animal will so consistently give the wrong answer, you can't help feeling he knows the answer full well and is being deliberately perverse—clearly a communication of some sort. Once during a game of fetch, Echo methodically brought me every item in the tank *except* the one I'd asked him to retrieve. Sure felt like teasing to me. (I felt the same way when I was a kid trying unsuccessfully to get one of my older brothers to pass me the potatoes at the dinner table.) A researcher elsewhere reported that one dolphin suddenly got everything wrong on a task it had performed virtually flawlessly many times before. It turned out that the fish, which the dolphin received via a mechanized dispenser, had gone bad. The dolphin, apparently, was signaling its distaste. Once the fish supply was changed, the dolphin's behavior returned to normal.

In trying to interpret "wrong" responses, it helps to have a good rapport with the animal, to know him and his tendencies, his ways of responding. You've got to remember that the dolphin is always communicating *something;* the question is what. Did he offer some alternative behavior? Simply stay in place staring at you? Swim off and ignore you? Start chuffing or tailslapping? Had he been working well until that point? Did something happen to set him off—a loud noise at tankside, for example? Was the behavior an old, familiar one or relatively new? Did you screw up the signal or somehow confuse him? Had someone new been working with him recently and perhaps given him confusing signals? Was he in a bad mood? Were you? All the dolphins seemed responsive to our moods and often would respond poorly if the trainer arrived in a funk.

A strong sense of empathy is, I believe, one of the principal traits of a good trainer. You need to be able to put yourself in the dolphin's place, to try to see the problem as he may be seeing it, and then be willing to change your approach accordingly. You need to be able to let go of your own ego. That often made the difference between good sessions and not so good ones, Michelle observed: if she wanted too badly for the dolphin to do whatever she was trying to train, she might fail to notice if he was having difficulty, or if he had figured out some other way to solve the problem. In fact, you may learn more from what the dolphin does "wrong"—then he's telling you who *he* is, not just showing you what you want to see.

A good trainer also needs to be quick and decisive. That was probably my major shortcoming. I don't think well on my feet. I'm inclined to mull things over, whereas training requires moment-to-moment interactions and on-the-spot decisions. I suspect I'm too self-conscious as well. A trainer has to be completely tuned in to the animal in front of her, not worrying about how she's doing or how she looks doing it.

Dedication is also essential; in fact, it may be carried to extremes in some truly top-notch trainers. The hours can be incredibly long, especially if you have a new or sick animal. Holly realized she couldn't be a trainer over the long haul because it wasn't her first priority. "I can't spend twelve hours a day for a month at one place and be sane or happy," she said. "I want to be doing something else sometimes—like sleeping."

Dolphins clearly respond differently to different trainers. For one thing, they seem to know what they can and can't get away with and with whom, who applies the strictest criteria for certain behaviors, and who will let them get away with something sloppier. Invariably when you first work with a given dolphin, the animal seems to test you, to put you through your paces to see what kind of trainer you are. And just as trainers often feel a greater affinity for one animal over another, the dolphins likewise seem to prefer some humans over others, responding more enthusiastically when a particular favorite works with them. Echo generally seemed to work better with Jill, for example, than he did with most of the rest of us. I felt more strongly connected to Misha; I believe the feeling was mutual.

In the end I came to view training sessions as a conversation of sorts, with both parties taking turns asking questions and making statements of various kinds. The trainer usually opens the conversation by bringing out the buckets and blowing the whistle. (The dolphin can sometimes start the discussion, however. One day, for example, Echo kept doing breaches and body slams between sessions, splashing huge amounts of water and creating quite a rumpus. He'd stop, however, whenever a crew member came over to play with him.) The dolphin can then engage in the conversation by coming over to the platform or, as Echo and Misha sometimes did, stay away, effectively not speaking to us. Generally we'd start by greeting the Boys, talking to them, touching them. That way both dolphin and trainer could check out each other's moods for the day. Then began the more formal exchange, the trainer asking questions by giving the signal for a certain behavior, the dolphin responding by performing the requested behavior—or by doing something else. If the dolphin didn't perform the desired behavior, whatever he did instead posed a question for the trainer, who then had to come up with a new response of her own.

Some conversations went very well, with dolphin and trainer achieving considerable mutual understanding. Others were much more difficult, replete with miscommunications or even disagreements over the desired topic of conversation. These "conversations" weren't exactly what I'd originally had in mind when I dreamed of talking with the dolphins, but they are as close to the dream as I'm likely to come in real life.

6

Senses and Sensibilities

*T*hese training-based "conversations" aren't enough, though. I still want a better sense of how dolphins perceive the world, of what goes on inside their heads. I felt that urge the first time I looked a dolphin in the eye at the Minnesota Zoo some fifteen years ago. I felt it so keenly with Josephine: What did she think? How did she see the world? With Echo and Misha the prospect became even more intriguing, since the world of humans was all so new to them. What did they make of it all? How did they fit our world into their worldview?

Again, this sense was much stronger than I've felt with any other animal. I do at times wonder what goes through our dog's mind as he plots his next escape from his pen, or what our cat is responding to when she tears around the bedroom pouncing on invisible prey. But that's more a passing curiosity. With dolphins the feeling is more akin to how I feel these days watching my infant daughter, Hannah. At four months her senses are still developing, and she perceives the world quite differently from the way I do. I'd love to know what she's registering as she gazes up at the hundred-foot redwoods that surround our home, for example, or what turns her from laughter one instant to screams the next. I look at her and am overpowered by a sense of wanting to know this person, to share her world, to "make common" as much as possible. Dolphins tend to evoke a similar response in me.

As I sat in the underwater viewing room one day watching the Boys between sessions, Misha came over to the window and stared back at me. I made faces at him. He came closer. I tapped my fingers on the glass. He touched it from the other side with his rostrum. We scrutinized each other for a good twenty minutes or more. I couldn't help but wonder what he was seeing as he looked at me. Did he recognize me as Carol, or was he simply responding to a person in the window? We can distinguish them by their dorsal fins, by other body markings; we respond to their "dolphinalities." What do they perceive about us as individuals?

I have little doubt that Misha knew who I was, at least when I was working with him during training sessions. I was less sure if he'd know me through the glass. Dolphins probably don't recognize us by our faces, at least not primarily. Facial recognition seems to be a particular forte of humans and some other primates. We even seem to have a particular area in our brains dedicated to it: neurologists have described a disorder, called prosopagnosia, in which a person loses the ability to recognize the faces of familiar people as a result of damage in a specific location.

Dr. Louis Herman and his colleagues at the University of Hawaii have studied a variety of cognitive abilities in dolphins. One of his students tested the ability of a bottlenose dolphin, Phoenix, to distinguish among human faces. In a "matching-to-sample" task (a basic procedure the animal was already familiar with), a trainer would briefly stick his or her head out of a hole in the wall, providing the "sample" human the dolphin was to learn. Then two humans would appear at tankside, and the dolphin had to choose the correct match—which person had the "sample" face it had just seen—and swim to that person. At first Phoenix was unable to distinguish among the faces unless the differences were accentuated somehow, by adding a wig, for example, or beard or mask to one. Her performance gradually improved, however, as she apparently learned what attributes to focus on.

Misha might have recognized me, in part, on the basis of my eyeglasses. I wore glasses with photogray lenses that usually turned quite dark when I was outside by the tank. So perhaps I was the one with the big, dark bug eyes. Or, since I often—but not always—took off my glasses during a training session, the one with the removable eyes.

I suspect dolphins may recognize each other visually more on the basis of larger-scale features—general shape, body markings, movement patterns, and the like—than on faces as such. They may well identify us in those terms, too. We often do so. I remember in high school, before I got glasses, I was able to recognize friends at some distance down the hall just from their overall shape and how they walked.

I'd be willing to bet, however, that dolphins recognize us more by our sounds than by our looks. Their hearing is especially keen, and I'd be most surprised if they couldn't readily tell one human from another based on voice quality, speech rhythms, and the like.

Dolphins certainly recognize each other by their sounds. Among bottlenose dolphins and some other species, each individual has its own distinctive "signature whistle," their version of a name. Echo's signature consisted of a series of three brief whistles, each one with an identical upsweep in pitch. Misha's was one longer whistle that swept up, down, and up again. If, for example, Misha calls out his own whistle pattern, it apparently means something like "Misha here." If Misha whistles Echo's signature pattern, it's like calling "Hey, Echo." If Misha's voice quavers as he whistles, it may translate along the lines of "Misha here and I'm scared."

Chances are dolphins can tell a fair amount by the quality of another's sounds, just as we can usually tell gender, some sense of age, emotional state, and even something about geographical origins from listening to someone speak.

The Boys may also have distinguished among us based on our touch, how we handled them—this one has steady, sure hands; that one is skittish, afraid of being bitten; this one likes to stroke; that one pats. Maybe we tasted different, too, when we put our hands in their mouths during morning inspections.

Most likely, they used some combination of all these things and more to distinguish among the many humans in their lives.

As I watched Misha through the glass, I tried to put myself in his place, to imagine the world through his senses:

Forget the sense of smell; as a dolphin you've lost it altogether—you don't even have an olfactory bulb in your head anymore. You've still got taste, but you're probably not as concerned about the taste of your food—you'll just swallow your fish whole, without chewing—as with tasting the waters for signs of the reproductive state of your schoolmates, for example.

You certainly still have touch, but no hands with which to grasp or caress; instead, you'll use virtually every part of your body—flippers, flukes, snout, teeth, even the penis—to stroke others, to explore objects, and to carry things around. You may spend as much as 30 percent of your time in physical contact with other dolphins, rubbing and caressing one another.

Your vision is good, but very different from what humans are accustomed to. Little or no color vision, for one thing. Not much binocular vision, either. Your two eyes function almost independently, and you can see nearly in a full circle around you. You've got the eyes of a prey animal—which you are. Remember, in the fully three-dimensional aquatic world, sharks can come at you from any direction—up, down, front, back, left, or right. It pays to have eyes not only in the back of your head but on the sides, top, and bottom as well. You have no eyelashes, no eyebrows. Your eyes glow in the dark even more than a cat's. They are built to deal with the darkness of the ocean depths and yet be able to withstand light levels just below the surface that may be several times the bright sunlight humans are accustomed to in air.

Perhaps most important, you now have what is undoubtedly the finest auditory system on this planet, covering nearly ten times the frequency range humans can hear. And in echolocation your hearing has been elaborated and adapted into a whole new sense, one that allows you to "see" with your ears, using sound to paint a three-dimensional picture of the world around you. You've even developed novel acoustic channels—you make sounds through your forehead and receive them through your jawbone.

Hearing is the most efficient long-range sense under water, and it clearly is of prime importance to dolphins; they make exquisite use of it in their echolocation. While echolocation is hearing based, it is a separate sense in its own right. Hearing and echolocation may be best regarded as two related and complementary senses, similar to our taste and smell. There even appear to be two separate sound pathways in the brain.

Ken Norris defines "echolocation" as the process whereby "an animal listens to the echo of its own sounds and then uses that echo to determine a bearing, range, or the characteristics of the echoing object." He further differentiates two types of signals used by the echolocating dolphin: orientation clicks, used for orienting to the general features of the environment, and discrimination clicks, used in situations requiring fine discrimination.

Navy sonar may be able to detect targets at greater distances, but an echolocating dolphin can gather much more elaborate information. The term "echolocation" is, in fact, something of a misnomer—dolphins do far more than just *locate* objects through sound.

By producing a series of "click" sounds and then listening to the returning echoes, the animal may gather information not only about the distance and direction of an object, but also about the object's speed, size, shape, texture, density, and even its internal structure. Various studies have found dolphins capable of feats we couldn't begin to match with our vision—discerning differences of only a few hundredths of an inch in the thickness of metal disks; distinguishing among visually identical cylinders made of different materials (bronze, aluminum, or steel); and detecting a one-inch steel sphere placed some three quarters the length of a football field in the distance. More important, at least as far as a dolphin is concerned, they can detect food fish or predators well before they are detected themselves.

Nobody even suspected that dolphins could echolocate until the middle of this century. For that matter, nobody knew for sure that bats could, either, till about 1940. As far back as 1793 an Italian monk speculated about "some new organ or sense" that allowed bats to fly in total darkness, but it took the discovery of ultrasonics—sounds above our hearing range—before scientists could finally establish that bats do indeed navigate by means of high-frequency sounds.

Researchers started to consider the same possibility in dolphins in the early 1950s, devising a series of experiments in which dolphins found fish or avoided obstacles at night or in extremely murky water. The problem was, the experimenters couldn't entirely eliminate the possibility that the dolphins were using keen night vision to perform these tasks. Bat researchers simply blinded their subjects to prove the point; dolphin researchers needed another approach.

In 1960 Ken Norris and his co-workers demonstrated unequivocally that dolphins echolocate. They managed to blindfold a bottlenose dolphin named Kathy and then get her to perform various tasks. Blindfolding a wet, smooth-skinned animal isn't an easy matter, and they tried any number of devices. The first model consisted of a pair of rubber falsies (junior-teen size) held in place by a steel spring over the top of the dolphin's head. Kathy didn't approve. In the end a soft latex suction cup over each eye did the trick. That continues to be the norm for echolocation research even now.

While blindfolded, Kathy successfully navigated an obstacle course in the tank. Hydrophones (underwater microphones) confirmed that she was emitting a series of clicks, sounding something like a creaky hinge, as she did so. No doubt about it—she was echolocating.

His work with Kathy also helped lead Ken to the rather startling conclusion that dolphins make sounds through their foreheads and receive them through their jaws, at least when it comes to echolocation.

While I was still a speech pathology student back in Minnesota, I convinced my professor in a class on hearing disorders to let me write a paper about hearing in dolphins. That was where I first ran into the name Ken Norris. I'd been reading about the controversy regarding how dolphins hear—or at least how sound is transmitted to the inner ear. In one study, done in pre–Marine Mammal Protection Act days, the researchers surgically removed various parts of the auditory system of several dolphins to see which portions made a difference in the animal's hearing ability. I could barely stand to read the paper (some of their conclusions are probably wrong, anyway). Then I read the passage in Ken's book *The Porpoise Watcher*, where he describes how he hit upon the jaw-hearing hypothesis. As he walked along the beach, he found a bleached dolphin skull in the sand. Examining the jawbone, he noticed that the jaw hollowed out toward the rear and the bone was extremely thin in that area—he could see light through it. In a living animal, the hollowed-out area of the jaw is filled with a special kind of fat, which attaches directly to the ear bone. He speculated that sound may come in through the thin area of bone (which he called the "pan bone" or "acoustic window") and be conducted via the fat to the ear. I was relieved to find a scientist who didn't take the "cut 'em up" approach, and much taken by the profound respect and enthusiasm for the animals conveyed throughout his book. I knew this was someone from whom and with whom I'd like to learn about dolphins.

I regret I've never had the chance to do field studies with Ken—I gather he's really in his element there. But in a sense he carries the field with him; he can turn the smallest observation into a nature lesson. Visiting him at his home atop a wooded plateau overlooking the Pacific Ocean—the "Ken and Phyllis Norris Experimental Gopher Ranch and Home for Wayward Animals," as the sign says—is a field trip in itself. The place is overrun by animals of all kinds: assorted dogs and cats, a flock of sheep that have been known to charge an unwary visitor, chickens and guinea fowl running through the yard, peacocks shrieking from the rooftop, an African gray parrot named Russ who can do a perfect imitation of a telephone ringing, followed by Phyllis's voice saying, "Kenny, it's for you."

I count among the great pleasures of working on this project the opportunity for free-ranging conversations with Ken. He is not only an extraordinary naturalist but a consummate storyteller as well. In fact, I think the two may go together—a naturalist observes an organism and then tries to piece together its life story. Ken has a knack for noticing things other people don't see, putting that together with a wealth of knowledge on the subject, and coming up with a good story about what is going on. Over the years he has proposed any number of what seemed to many to be preposterous notions about dolphin sound production and reception. Over the years further study has tended to bear out his notions more often than not. There is now overwhelming evidence to support his contention that a dolphin's sound source is located in the forehead, not in the larynx as it is for other mammals (though the precise location is still a matter of some debate), and that sound enters the ear via the "pan bone" of the jaw.

It's hard to know just what sort of image dolphins construct from echolocation, what sort of "picture" they derive from sound. We humans tend to think in terms of vision and are likely to come up with a visual construct, something like the output of a modern ultrasound machine. I like to think of echolocation more as a long-distance form of touch—dolphins can reach out and touch the world around them with sound. In fact, hearing and touch are closely related senses. Both are based on picking up vibrations; it's just that hearing developed to deal with much faster vibrations and ones that are transmitted through the air. For us touch is a close-up sense (as is taste); hearing is a distance sense (as is vision). Echolocation, in a way, is both—the echoes come back from a distance, but the dolphin first contacts the object with its sound.

Furthermore, given the greater density of water, dolphins may be able to *feel* the sounds on their bodies. The "feel of sound" may be a part of their echolocation experience.

This tactile quality of sound under water may have a couple of provocative secondary effects as well: at very high-intensity levels, dolphin sounds may function as a weapon, at low levels as a caress. Another of Ken's seemingly outlandish conjectures, his "big bang" hypothesis, suggests that dolphins and other odontocetes just might be able to stun their prey with sound. He noted that many species of odontocetes have few or no functional teeth and yet can catch small, fast-moving, maneuverable prey. The sperm whale, for example, which feeds on swift and slippery squid, has teeth only in the lower jaw, and even those don't erupt until puberty. Also, whalers sometimes found squid still alive and with no trace of teeth marks inside the whales' stomachs. One possible explanation is that the whale first zaps the squid with its sonic "stun gun," then slurps up its meal at its leisure. This technique apparently is used on a smaller scale by at least one other kind of animal, the pistol shrimp. Dolphins are capable of producing sounds at energy levels sufficient to kill or debilitate fish, and researchers have seen fish schools lose their orientation and turn lethargic around feeding dolphins. The evidence remains largely circumstantial, however. No one has yet witnessed the process in action, and laboratory studies have thus far been unable to confirm the notion.

Low-intensity sounds may be a whole other story. After Jo and Arrow first arrived at the DRC in Florida, Michelle and I swam with them several times. It was the first time I'd been in the water with a dolphin without wearing a wet suit. Jo scanned me thoroughly, and I was struck by how much I could feel the echolocation pulses—a tingling sensation that seemed to run up my spine. Surely dolphins must feel it, too.

They probably not only feel the sound but have added it to their remarkably varied and extensive repertoire of sexual behaviors. One commonly observed dolphin activity is what's referred to as "beak to genital propulsion," in which one animal sticks its snout into another's genital slit and gives its partner a ride. Males as well as females have genital slits, so both genders can participate, and with members of either gender. If, as some trainers have observed, the one doing the propelling were to echolocate softly in the process, it could well give the propellee a nice buzz. Dolphins clearly use sound with great facility and in all manner of ways, perhaps even as a built-in vibrator. Very handy for giving a buddy a "buzz job" now and then.

7

*E*cholocation
and *E*yecups

*T*he arrival of the Lags had set back both our general training and our echolocation research with Echo and Misha. By January of 1989, once all the Lags were in place and eating properly, life settled down enough that we could at least resume a regular training regimen with the Boys. The research itself, however, would require the large tank, to which the Lags had sole claim for the time being. Meanwhile, we started some preliminary training for the echolocation study, as well as continuing work on husbandry behaviors and "fun stuff."

Ken somehow couldn't bear to watch us train the dolphins while lying on our bellies, hanging over the edge of wooden platforms. That just looked too uncomfortable to him. He built us a contraption designed to be easier on his mind and our bodies. It consisted of a large blue plastic barrel, weighted with cement in the bottom to submerge it, attached to a framework of metal pipes that fit over the rim of the tank to hold the whole thing in place. The top of the barrel was at water level, so the barrel filled with water. That meant we had to wear hip-waders of the sort fishermen wear in order to stay dry. That way we could stand upright within the barrel, our arms at just the right level to work with the dolphins. It seemed like a good idea, but the thing never caught on. Not with us, not with the Boys. The cement made it incredibly heavy and nearly impossible to move; the pipes dug into and ate away at the concrete of the tank side. Holly said she felt as though she were in one of those wire hoops used to hold Easter eggs when you dip them into the dye, so we dubbed Ken's creation the "trainer dipper." Echo and Misha wouldn't even come near it at first. They didn't get much better with time. Whatever advantage we gained by being at their level was lost if they stayed at arm's length. Maybe the Boys didn't want us at their level. We retired the dipper after a two-month tryout in the thirty-foot pool.

That was just one of several constructions Ken devised for the dolphin project over the years, which met with varying degrees of success. He has a noted propensity for collecting scraps, odds and ends, and assorted gewgaws, from which he may someday build something. He built a wondrous, if perennially unfinished, tree house on his land, made in part from the hull of an old fishing dory he salvaged from somewhere.

Not long after the demise of the "trainer dipper," Ken erected a small wooden shed on the deck next to the pool. He intended it as a sort of a researchers' blind. The researchers (mostly that meant me) could sit inside, unseen by the dolphins. We could also keep the experimental equipment and papers in there, away from the elements. The shack reminded someone of the stall where Lucy sits as she counsels Charlie Brown in *Peanuts* cartoons. It promptly became the "psychiatrist's booth." I felt more comfortable and had a better view of the session simply sitting at tankside. The electronics equipment was safer down below, in the underwater viewing room, and was far too unwieldy to haul up top for each session. We did use the booth, however—but not quite in the way Ken had in mind. It made a great storage shed for the dolphins' toys and other tankside gear. Dawn used it as a sort of reconnaissance tower for her Lag study, setting up her video camera on top of the shack whenever she did twenty-four-hour observations; the higher perch provided a good angle from which to get the entire tank in the picture at once.

Ken's talents are hardly limited to research and construction projects, however. Among other things he's also a mail-order minister—Church of Gospel Ministry. Randy and Michelle got married in early March, with Ken performing the ceremony (as he would do later for Joel and me). His marriages *are* legal, despite the dubious nature of his credentials. He does them only for people he knows, mostly students or former students, and seems each time to get even more nervous than the bride and groom. (In part, I gather, because he's generally good for one interesting gaffe per wedding. In our case he first asked me if I would take Joel as my wife. Whatever.) His services are splendid, however, offering his unique naturalist's-eye view both of the couple getting married and of life more broadly. I thought Randy and Michelle should hold their wedding at the lab with Echo and Misha as "best dolphins," but they decided otherwise. Weather in March is iffy at best around here, and the deck around the pool wasn't well suited to dancing.

A couple days before the wedding Michelle took Randy's parents out to the lab to show off her dolphins. As a special treat she took Randy's father in swimming with the Boys. Misha bit him. Just a small nip on the hand, but it did draw blood. Not quite the impression Michelle had hoped to make on her father-in-law-to-be. We were somewhat surprised that Misha did the biting; Echo was usually the nippy one. Maybe Misha wanted to be best dolphin, after all. More likely, he was feeling crowded in the thirty-foot pool with a stranger. Perhaps it made a difference, too, that the stranger was male. By and large, Echo and Misha dealt mostly with women. Howard was the only man who interacted with them on a daily basis. Generally, if there were more men around, especially men in the tank, that meant something was up—like one of their quarterly physical exams.

By the end of March we started gating the Lags so Echo and Misha could have access to the fifty-foot tank again. We settled into a routine in which the Boys got the big pool roughly half the week, the Lags the other half. At last we could begin more serious work on preparing Echo and Misha for the echolocation study.

Basically, we wanted to look at some of the collateral behaviors associated with echolocation—what does a dolphin *do* as it echolocates? An echolocating dolphin, after all, isn't a stationary sonar device; it's a highly mobile, actively scanning, living, breathing organism. As the animal echolocates, it's also swimming. In addition, it may be swinging its head to and fro, rolling on one side, opening its mouth, or moving in any number of other ways, overt or subtle. It may even be shifting its sound beam internally, within the forehead.

Most echolocation studies, after the early few, have specifically sought to eliminate such behavior. This is usually accomplished by having the dolphin rest its rostrum on a chin cup or bite down on a specially designed plate so the animal will remain stationary. The procedure makes good sense from a physics standpoint—you can locate the sound source precisely and make more sophisticated measurements.

We wanted to work with free-swimming animals, however. We figured these movement patterns might be a vital component of echolocation—that to ignore them would be a bit like trying to understand ballet by studying only the music, without regard to the dance.

One reason for believing that the "dance" of echolocation may be important is that the "music" is highly structured and highly directional. Whereas most communication sounds—whether ours, a dolphin's, or another species of animal's—are broadcast more or less omnidirectionally, a dolphin's echolocation sounds are projected outward through the fatty melon in a tightly focused emission beam. The highest frequencies and amplitudes form a very narrow central beam that follows the line of the rostrum but at a slight angle above it; both frequency and amplitude drop off sharply farther from center.

The structured nature of the outgoing beam means that as the dolphin moves, so does the sound beam. Imagine wearing a flashlight on your forehead. As you walk, the light beam bounces a bit with each step. If you move your head from side to side, the light moves with it. If you yawn or stretch or wiggle your ears, that too could affect the beam. If you're wandering through the darkness, you may want to move your light beam in all manner of ways, from large-scale sweeps to tiny circles, as you explore the world around you.

Or, to return to my favorite analogy, the dolphin's movements could serve to play the sound beam across an item of interest somewhat as we might investigate an unfamiliar object by running our fingers over its surface. In fact, there may be some interesting parallels between echolocation and what psychologist James J. Gibson calls "active touch," which he describes as "an exploratory sense rather than a merely receptive sense." Echolocation clearly fits that description as well. Gibson looked at how well people could match pairs of unfamiliar cookie-cutter-type shapes purely based on touch. Subjects were nearly twice as accurate if they were allowed to make active use of their hands in exploring an object, as opposed to merely having the object placed in their hand. We figured an actively scanning dolphin should likewise be a more effective echolocater than a restrained one.

Our procedures called for blindfolding Echo and Misha, using the same sort of latex eyecups Ken had originally developed for Kathy thirty years earlier, to ensure that they would echolocate. Then we would ask them to swim from the training station at one end of the large pool to perform one of several tasks in front of the underwater viewing window at the opposite end of the tank. Tasks included locating a stationary target and touching it with their rostrums; locating and retrieving a plastic ring, either one floating at the surface or one sinking down through the water column; and locating and eating a dead fish drifting in the water.

The stationary target consisted of a plastic pool float attached to the end of a metal rod. A wooden framework held it three feet out from the viewing window and three feet under water to reduce the amount of sound bouncing back from the window and the surface. The rings were made of plastic tubing taped into a circle of approximately nine-inch diameter. One ring had only air in it so it would float; the other ring had enough water in the tubing to make it sink slowly. Fish were herring and capelin, the Boys' usual food items. We tossed the rings and fish into the water with a splash on some trials, making a sound the dolphins could orient to initially; on other trials we placed them quietly in the water using a long-handled clamp. The target was always in the same location, whereas the exact position of rings and fish varied from trial to trial.

We recorded the dolphins' movements during trials on two video cameras in the underwater viewing room. One camera behind the target window provided a head-on view of the animal as it approached and performed the task. A second camera at one of the side windows gave a lateral view of the dolphin as it swam across the tank. The two cameras were linked into a split-screen image so we could monitor both views simultaneously.

We suspended a set of "guide poles" in the water behind the dolphins' swim route (as seen from the side-view camera) to serve as a giant measuring stick of sorts, allowing us to gauge each dolphin's movement toward the target and his depth in the water column. This consisted of a string of ten ten-foot lengths of PVC pipe, set a yard apart from one another, each marked with black-and-white bands at regular intervals. A "spiderweb" circular grid system, drawn on clear acetate, in front of the target window (head-on view) helped us measure the dolphins' head movements. We picked up their sounds via an array of four underwater microphones, called hydrophones, one placed directly on the target itself and the other three located on a circle around the target hydrophone.

Training focused both on teaching Echo and Misha the tasks that would be required of them and on getting them to wear the eyecups comfortably. The tasks were the easier part. Basically every aspect of the training required successive approximations, working up toward the final goal and adding new elements slowly.

The stationary-target task took a while to train. We'd already started work on that while the Boys were still in the thirty-foot pool, getting them to touch a hand-held target. First the trainer held the target. Then another person took it and moved progressively farther away from the training station so the dolphins would have to swim off and find the target. Initially we splashed the target on the water's surface to get their attention and to give them an auditory cue as to the target's location. Subsequently we used a "clanger," hitting two pieces of metal together under water to signal them they were to go find the target. We also started moving the target itself under water, gradually working toward the three-foot depth we wanted. Within a month or two both Echo and Misha were swimming the length of the pool, unblindfolded, and touching the target on its pole beneath the water.

Allowing himself to be blindfolded requires considerable trust on a dolphin's part. Some take to it far more readily than others; some never accept it at all. We tried to ease Echo and Misha into the eyecups gradually. First we simply showed them the eyecups, letting the Boys see and grow accustomed to them as objects. Next we'd touch the eyecups to our own bodies, and then touch them very briefly to various parts of the dolphins, letting them feel the latex against their skin. Simultaneously, we worked on holding a hand in front of one of their eyes, so they'd learn to accept having their vision blocked briefly, before we tried it with the eyecups themselves. Then we simply held an eyecup over one of their eyes, without putting any pressure on it. Often we'd first place the eyecup over one of our own eyes to show them what we had in mind. From there we started sticking the eyecups to their melons and other parts of their bodies, so they could feel the suction initially on some less sensitive parts of their bodies. We didn't actually try placing an eyecup over one eye with suction applied until they'd come to tolerate all the earlier steps.

Both Echo and Misha reacted even more strongly to removal of the eyecups than to their placement in the first place; I suspect the release of suction felt especially strange. I tried out the eyecups on myself to see what it felt like, though I couldn't get one to stick over my eye; you need a fairly smooth, flat surface, so the best I could do was test one on my stomach.

Once they accepted eyecup plus suction, we started yet another progression. First an eyecup over just one eye. Then an eyecup on one eye and a second eyecup on the melon or somewhere else, giving them the feel of two eyecups but still allowing them some vision. We'd also place one eyecup over an eye and hold a hand over the other eye, blocking their vision but without making them wear both eyecups or go anywhere while blindfolded.

As we worked on the eyecup progression, we also gradually introduced the various other aspects of the experimental setup—the "guide poles," the "spiderweb," hydrophones, cameras, people in the viewing room and on the deck—to give Echo and Misha time to adjust to each new element. The Boys fled when we first tried putting the guide poles in the water. To them, apparently, it was as though we'd thrown a net into the water. We took it out and started over, removing all but one of the hanging poles and then adding from there. We also shoved the poles off to the side of the tank and then moved them progressively closer to the desired position. Echo and Misha did come to accept the pole setup, though they generally preferred to keep their distance.

They had to contend with clatter in the viewing room, plus the occasional whines, buzzes, and squeals from the electronic equipment as well. Sometimes a particularly loud or irksome noise seemed to set them off, and they'd go on strike, refusing to participate for the rest of the session.

By midsummer we were ready to start recording actual data sessions. At least we thought we were; the Boys had other ideas. We set up to record our first real session on July 27, 1989. We'd send one of them on a run with no eyecups. No problem. Then with one eyecup. Again no problem. Finally, both eyecups. Problem. Neither Echo nor Misha wanted to go anywhere while blindfolded. They'd just stay by the training station or, at best, swim a short distance toward the target and turn around.

Once he did start moving, Echo, ever the clever one, reacted by swimming to the bottom and rubbing the eyecups off against the tank floor. We called a time-out whenever he did that to break him of the habit. Misha, poor Misha, just hovered in one place, looking befuddled. He'd sway a bit, moving his head from side to side in what we came to think of as his Stevie Wonder imitation. Eventually, both devised a variety of techniques for getting the eyecups off when they wanted. Misha later picked up the trick of rubbing them off on the bottom. He also learned that he could sometimes pop one off by opening his mouth and moving his jaw as he swam. Echo discovered that if he swam really fast, one or both eyecups sometimes slid back off his eyes. This created something of a dilemma for us. First, we couldn't necessarily tell if he was deliberately trying to remove the eyecups and therefore we should time him out, or if he was simply swimming fast because that was his wont. Second, while this maneuver would sometimes cause an eyecup to slide off his eye entirely, even up onto his back, at other times it only subtly displaced the cup, so we couldn't tell whether his eye was still covered or not—he might look fully blindfolded but actually be "peeking."

The biggest problem, however, was that they simply weren't echolocating, blindfolded or not. Far from turning on their echolocation, that marvelous special sense, as we'd expected, they remained utterly silent. Always full of surprises, these Boys.

I'm told trainers and researchers elsewhere have experienced similar reactions. In retrospect it makes a certain amount of sense. Dolphins often go quiet when in strange or frightening circumstances. If someone blindfolded me, I suspect I'd be reluctant to make any noise, too; why advertise my whereabouts to potential predators, whom I wouldn't be able to see? Echolocation functions in front of and slightly above a dolphin. Sharks tend to attack from below. Echolocation may be a dandy early-warning system at a distance, but it's not going to tell you if something is sneaking up on your tail. For that you want those 360-degree eyes. So first off, Echo and Misha needed to feel safe enough to start making noise when they couldn't see.

Furthermore, they didn't necessarily know that we wanted them to echolocate—knowing what the humans wanted was always a guessing game for them, I think. They were no longer echolocating much on their own. That, too, makes sense. Once they'd learned the layout of the tank, there wasn't a whole lot to echolocate on. Besides, the tank water was generally relatively clear, unlike the murky waters back home, and the distances not great, so vision alone worked just fine. One of Michelle's big regrets, in retrospect, was that she didn't put their echolocation on a cue from the outset, so she could have them echolocate on command.

We had to backtrack, to do remedial echolocation training. We tried everything we could think of to get them echolocating again. We put all sorts of novel objects in the tank, hoping they would want to check them out closely. Whatever checking out they did, they did visually. We tried live fish. We even tried putting live fish inside one of those clear plastic balls used to exercise pet hamsters. Echo and Misha could see the fish but not get at them, which we figured should entice them to explore the situation more closely. Nary a click.

Michelle worked on simply getting them to make noise on cue. She'd already "captured" that behavior much earlier. While one of them stationed with his rostrum on her hand, she'd tap on the area around his blowhole. The Boys generally responded with a sputter or some other sort of noise. Michelle then blew the whistle, thereby reinforcing them for making sounds when she touched their blowholes. They had both caught on fairly quickly, so we could get them to vocalize on cue. The sounds were hardly echolocation clicks—more of a wet-sounding blat or raspberry—but at least it was something. They had learned that behavior with their heads above water, however, so Michelle worked on getting them to make the sounds under water. Then she put eyecups on them and asked them to vocalize. Next she had someone else move a target or another object in front of them as they did so, trying to give them the idea it was all right to make sounds at an object under water. They did squawk and blat some, but they still weren't echolocating.

In late August, Ken proposed a new approach. He decided we should make the tank water murky enough so Echo and Misha couldn't rely on their vision. We wanted something nontoxic, nonpolluting. Milk. That should do the trick quite nicely, Ken suggested. Michelle worried that it might lead to bacterial growth in the filter system, however. We got around the problem by shutting down the filtration system during the experiment, moving the Boys out afterward, and then draining and refilling the tank. Bacteria levels in the tank were no higher afterward than before.

On August 22, for the two P.M. session, we poured four gallons of milk into the water of the thirty-foot pool with the Boys. Undoubtedly the first dolphin milk bath ever. It's amazing how murky just a little bit of milk can make thirty-five thousand gallons of water. Both Echo and Misha responded by breaking away from their training stations, sinking to the bottom of the tank, and letting out a great big bubble from their blowholes, something they generally seemed to do when seriously perplexed. (I always thought of those bubbles as a giant question mark, or as "What the . . . ? ? ?") We threw fish into the water for them, but they went after only those that landed very close to them; they wouldn't echolocate to find any.

Both were reluctant to return to their trainers and do any work. Misha refused to do much of anything. Echo finally returned to the platform, and Michelle did some targeting work with him. At the end of the session he produced three or four very brief click trains in response to the target. Michelle didn't feel she was able to catch and reinforce the echolocation sufficiently at the time, however. They didn't suddenly "click in" and start echolocating, at any rate. The milk experiment wasn't the turning point we'd hoped for, but it may have helped Echo get the idea.

Since the milk wasn't a howling success, our next move was to try a night session. The following day we held a session with the Boys in the big tank at nine P.M. It was a dark, moonless night, so their vision should at least have been restricted. They tricked us again, however. Apparently they were able to find and touch the target using just their eyesight—again, they can see far better than we can in the dark. Or their hearing is so keen, they may have been able to detect the targets just by listening, perhaps to small air bubbles released by the target or the sound of the rings moving through the water, without producing any clicks of their own. Echo didn't make any sound this time, except a few frustrated-sounding blats. Misha, however, gave a few brief echolocation trains, though only *after* he had touched the target and was on his way back to the training station. Not exactly a resounding success, but again, maybe it helped.

They still didn't quite have the idea of what we wanted, though. Michelle continued to work with them on making sounds—any sounds—as they moved toward a target. Sometimes those sounds included a few clicks, but not the stream of click trains we were expecting. We also progressively increased the distance the dolphins had to swim to reach the target. Misha still refused to go anywhere with eyecups on. Echo sometimes managed to find his way to the target, at least when it was relatively close, without echolocation—whether simply by listening or by a process of stumbling around in the dark, we were never quite sure. I suspect they knew their tank well enough by then that they could navigate it fairly well without sight—just as we can find our way to the light switch in a dark room if we know the room well.

Echo finally put it all together about a month later. We didn't notice any great moment of revelation, no "aha" experience. He'd gradually increased the number of clicks he produced in response to Michelle's "make noise" signal. In the process, I imagine, he discovered that nothing bad happened when he clicked, and since it made getting around easier, well, why not? He made his first official data run—both eyecups, all the equipment in place, target at its ultimate location, and click trains—on September 29. We, of course, all jumped up and down, shrieking and creating a general hullabaloo. Maybe that's when the "aha" came. When he got back to the station for his reward, Echo looked at Michelle as if to say, "That's it? That's all you wanted?" Misha still didn't get it.

We concentrated our efforts mostly on Echo over the next couple months, since we were able to collect actual data from him. We continued to work with Misha, however, trying to provide whatever additional reassurance, coaxing, or enticement he might need to complete the task. We tried, among other things, sending him on runs with Echo, figuring he might learn the task through imitation. No. No copycat he.

Once Echo had clearly caught on to the stationary-target task, we introduced ring-retrieval trials. That turned out to be an easy transition, since retrieving rings was already a part of the Boys' "fetch" repertoire. When he proved successful with both tasks, target and ring, we started mixing them up so he wouldn't necessarily know which he'd be asked to do on a given trial. This meant he'd be less likely to fall into stereotyped movement and sound patterns.

Somewhat later we added fish trials to the list. That required no specific training, other than first sending him on a few fish runs without eyecups so that he knew there was now a third possibility. Echo was more than willing to find fish and eat them. Unfortunately, so was Misha—though not while wearing eyecups himself. No, he waited until we'd blindfolded Echo for a fish run. When he heard the splash of the fish on the water, Misha would break away from his trainer and snatch the prize away from Echo. Time-outs didn't much discourage Misha, who seemed to relish beating his tankmate to the punch.

We decided instead to eliminate the splash. We fashioned a crude device from a plastic clip, the sort used to keep potato chip bags closed, taped to the end of a long pole. (We were not, by and large, a high-tech operation.) Then we could clip the fish by the tail and place it quietly in the water. That seemed to stop Misha, but we ran into a new problem. We'd had Echo practice ahead of time pulling a fish out of a hand-held clip. Easy. Not so easy, though, when he was blindfolded and the water current was turning the fish this way and that. On one of his very first fish-'n'-clip trials, Echo simply couldn't get the fish out of the clip. He clearly knew where it was and went for it. The fish was a capelin, small and somewhat curled to begin with, and the current forced it up behind the clip. After a brief struggle Echo gave up and headed back toward the training station. He'd gone only a short distance, though, before he stopped, swam back and chomped down full force on one of the hydrophone cables. He left little doubt about what he was communicating that time. We quickly added a release lever to the clip, so we could release the fish under water as soon as we knew that Echo was on his way—and that Misha wasn't.

The fish task posed one additional problem we weren't able to solve: Echo thereafter seemed less interested in the target and ring trials. He started to balk more often, swimming about halfway across the tank and then turning back without completing the task. My bet was that he scanned the tank early in the run, and if he ascertained that there was no fish out there he decided, "Why bother?"

That period also brought some major personnel changes. October marked the end of Dawn's one year of training-free Lag observations. Shedd Aquarium hired Lisa Takaki, who'd been working as a dolphin trainer at Busch Gardens in Florida, to begin training the Lags, preparing them for the lives they would lead once Shedd's new facility opened.

Lisa, whose long raven ponytail stood out among our predominately fair-haired crew, was the first person I'd ever met who really had run away and joined the circus when she was younger. She'd done both trapeze and high-wire acts, and her grace and agility still showed as she worked around the lab—though even she fell into the tank once while vacuuming, making the rest of us feel better. She'd performed in ice-skating shows as well before finally finding her true calling as a dolphin trainer. She is undoubtedly one of the finest trainers I've ever seen, and the rest of us learned a great deal from her. The animals seemed to respond to her eagerly, joyously; she, in turn, seemed acutely tuned in to them, keenly aware of every action and reaction.

On the down side, we had to bid farewell to Randy and Michelle. Our grant money was running out, leaving the project in serious financial trouble. There was no longer sufficient funding for Randy's position as project coordinator, and Michelle's would be cut to part-time. The professorial position Randy had hoped for at UCSC didn't come through. (He is still considered an adjunct professor and can supervise graduate students; he just doesn't get paid for it.) Much as they hated to leave Santa Cruz, Echo and Misha, and the project, they needed to make a living. Brookfield Zoo in Chicago offered Randy a job, one created especially for him in their Conservation Biology Department, which would allow him, among other things, to continue his study of the Sarasota dolphins. They hired Michelle, too, as a dolphin trainer at the zoo. The opportunity was too good to pass up.

The two of them left in late October, amid some serious grieving all around. They would still coordinate Echo and Misha's release, but they would no longer provide the day-to-day leadership. Holly took over, with some trepidation, as head trainer for the Boys. She was good with both the people and the animals, but the responsibilities were heavy, and she was still new to the business.

For reasons that had nothing to do with either the dolphins or the crew, that autumn is forever fixed in my mind as the "Fall from Hell." It started off with acute appendicitis and an emergency appendectomy in late August. I was just out of the hospital when we did the milk experiment with the Boys; I showed up to watch but wasn't of much help. I missed the night session altogether.

Then on October 17 at 5:04 P.M., just as the country was settling in to watch the first game of the World Series between San Francisco and Oakland, the earth shook as I'd never felt it shake before. The great quake of 1989 registered 7.1 on the Richter scale. Most of the national media called it the San Francisco earthquake, but its epicenter was actually in Santa Cruz. I was up on campus at the time, in the middle of a bridge, which I fled as I saw the concrete start to ripple and the glass globes from the street lamps come crashing to the ground. As I made my way home, I discovered the quake was far worse than I'd first realized. Most of the bridges across the San Lorenzo River downtown were closed. The whole of downtown—what had once been the lovely Pacific Garden Mall—was virtually demolished. Three people died there, two elsewhere in Santa Cruz. Given the amount of damage in the area, we were fortunate there weren't even more deaths.

I found my home intact, with only minor breakage, but with no power and no phone. I wasn't able to get through to friends locally, much less to worried friends and family back east, until the following night. Nor was I able to get through to the lab to see how the crew and the dolphins had fared. I had visions of the tank floor cracking open, or the underwater windows breaking and all the water rushing out, leaving the Boys and the Lags stranded in a couple feet of water.

I made it out there the next day and was relieved to find that all the people and all the dolphins were fine. The tank complex, designed with earthquakes in mind, had held up quite well. The pools shook a lot, apparently, losing a good deal of water in the process, and a couple of the underwater viewing windows leaked a bit more than usual afterward, but they suffered no major damage.

That's not to say the dolphins were the least bit happy about this latest turn of events, however. They remained edgy and jumpy for weeks, as did the crew. We continued to get jolted by aftershocks, some of them quite substantial earthquakes in their own right. The emotional aftershocks lasted even longer. It took months, I think, before people stopped jumping up and heading for the door whenever a big truck went by, anything that set the ground to rumbling in the slightest.

The Boys may even have sensed something ahead of time. Holly, who had worked with them that day, said they had been notably weird, especially Misha. In the training record book for the two o'clock session, three hours before the quake, she'd noted that "he just wasn't himself"; she'd even gone so far as to declare him "brain dead." I wish I knew if they realized the earthquake was a natural phenomenon, or if they considered it just one more crazy thing the humans were doing. Maybe they held us responsible: I could almost hear Echo's demand, "Stop rattling the tank, will you?" and Misha's plea, "What do you want from us now?"

The crew had been in our office trailer overlooking the tank as the shaking started. Lisa, the new Lag trainer, had just arrived a few days earlier from Florida. When she asked, "What's that?" our quake-seasoned crew calmly responded, "It's an earthquake—welcome to California." But as the shaking gathered momentum, they all fled the trailer in terror—led by Weasel, Jill's aged and ailing dog, who hadn't moved so fast in years.

I learned that Casey's home, a spectacular wood-and-glass house set amid the redwoods of the Santa Cruz mountains, not far from the epicenter, had been destroyed. One entire three-story wall simply collapsed. The house Jill had been living in, next door to Casey's, was also seriously damaged and, at least temporarily, unlivable. Dawn, too, lost her home to the quake.

That fall also brought some far smaller-scale but still intense personal trauma for me. I faced turning forty at the end of November, an unsettling prospect at best. As things turned out, however, I didn't have time to worry about that. The specter of middle age was greatly overshadowed by the sheer terror of my impending "orals." I had scheduled my preliminary oral exams for the Ph.D. degree for November 27, three days after my birthday.

Orals represented the granddaddy of all my fears. Talking has never been my strong suit; having to speak to groups or authority figures tended to elicit a full flight response; standing in front of a panel of four professors who would be grilling me about what I knew and didn't know—and deciding whether I could continue in the program—loomed as something monstrously, unimaginably horrifying. Years earlier, when I'd finished my bachelor's degree, I'd considered going on for a Ph.D. in psychology but decided against it, largely because being a doctoral student would mean being a teaching assistant; that in turn meant talking in front of groups of people. I was simply unwilling to face that prospect back then. I hadn't even considered orals. I suppose it was some measure of how far I'd come in the intervening years that I had worked as a teaching assistant several quarters already (not without getting a stomachache the night before I had to get in front of a class, however), and that I'd gone as far as to schedule orals and hadn't yet arranged to skip town.

I spent much of my time during that fall—time I would rather have spent at the lab watching Echo and Misha's progress—studying for and fretting about orals. The appointed time came. I'd sooner have faced another appendectomy. I'd have to rate the experience as four of the most miserable hours of my life. I answered the questions, after a fashion, but defensively, as though fending off an attack with each one. My mind seemed to function at slowest speed, hopelessly mired in uncertainty, and the more I struggled, the deeper I sank. All I'd hoped for beforehand was just one good, meaty question that I could answer fully, articulately, just one question I could say afterward, "At least I answered that one well." I didn't get it. Not a one came out right.

I waited in the hallway for nearly an hour as the committee decided my fate. I passed. I was greatly relieved—I'm not sure what I would have done if I'd had to go through it a second time—but I couldn't muster much in the way of joy. Maybe my answers hadn't been quite as bad as I'd feared, but my style certainly left a great deal to be desired. Often an orals committee will pass somebody but recommend further course work in a particular area. In my case they recommended a public speaking class.

I went out to the lab the following day. I needed a dolphin fix to remind myself why I had chosen this business in the first place. We ran an experimental session that afternoon, focusing on Misha. This time he got it. He completed his very first successful two-eyecup trial. Not only that, he was so excited about his accomplishment, he could hardly contain himself. Far from Echo's cool "That's it?" response his first time out, Misha's attitude was more "Ooh, ooh, I did it!" He practically danced around the tank, and his built-in dolphin grin seemed wider than ever. Bless you, Misha. He took my mind off orals and then some. As far as I was concerned, the Fall from Hell was over.

Once Misha finally got the idea, his performance was stellar. He may have been slower to get started and more tentative in his approach to things than Echo, but once he got going, he was far more consistent. Echo might respond beautifully one time and not at all the next. We concentrated primarily on Misha over the ensuing three weeks, and by Christmas we had all the basic echolocation runs we needed of all types—stationary target, ring retrieval, and fish—from both dolphins.

8

It Don't Mean a Thing If It Ain't Got That Swing

By Christmas we had recorded more than four hundred success-ful two-eyecups trials from the Boys (not all of them analyzable, however, for reasons including poor tank visibility on some occa-sions and equipment malfunctions). About two thirds of the runs were from Echo, since he'd started earlier.

I started by trying to characterize their swimming patterns—bearing in mind, of course, the limitations imposed by tank size. Both Echo and Misha tended to take a breath at the outset of a run toward the target, then swim at midwater level till they reached it. Both animals generally beat their flukes two or three times at the outset, followed by a brief glide with the body held in a straight line parallel to the tank bottom and the pectoral fins held out like wings. Toward the end of the run they would bring their flukes down sharply in a sort of braking maneuver; often they also used both flukes and pecs to steer themselves into position to touch, retrieve, or eat the target item. Both dolphins also quite consistently rolled onto their left side when they touched the target; they sometimes did so on ring and fish trials as well. Occasionally one of them would swim most of the run on his side, but generally they were upright until they neared the target.

Because we intermixed several different trial types, and the dolphins didn't necessarily know which was coming next, swimming behavior didn't appear to become as stereotyped as it might otherwise. The stationary-target task elicited the most consistent swimming patterns—not surprising, since it is the most predictable task, with the target always in the same location.

Several previous investigators observed that an echolocating dolphin sometimes moves its head back and forth, or in subcircular fashion, as it closes in on a target. Our results support the notion that this head scanning is more pronounced on more difficult tasks. In general the stationary-target task elicited less scanning than did the ring-retrieval or fish trials—that is, trials involving a moving target. On target runs the Boys were more likely to do a quick, short back-and-forth scan about halfway down the pool from the target. Ring and fish trials were more likely to bring longer and larger scanning motions, often in a more circular pattern, when the dolphin was much closer to the object in question. Furthermore, a ring or fish floating at the surface seemed to pose a more difficult problem, presumably because of sound bouncing back off the water's surface, and hence to elicit more head scanning than did the ring or fish drifting down through the water column.

In the course of the study we noted that both Echo and Misha sometimes opened their mouths as they swam toward the target. Since a dolphin's sound-reception system, at least for high-frequency echolocation sounds, involves the fat-filled jaw canal, this behavior could actually be a listening technique—and one that could be particularly useful if what you're after is a fish that you want to catch in your mouth. Given the shape of the dolphin's jaw, it could even serve as a sort of megaphone for the incoming sounds. Misha also tended to open his mouth as one of his eyecup-popping techniques, however, making for some confusion on this score.

A number of authors have pointed to the keen hearing ability of dolphins and suggested that they probably get a great deal of information simply by listening—or what has variously been called "passive acoustics," "passive sonar," and "passive hearing"—without generating any sound of their own. One facility reported that a blindfolded dolphin followed a live fish all around its tank simply by listening, without producing any sound at all. Such listening behavior may be far more important than we've realized. Echolocation is a high-energy activity, probably involving considerable wear and tear on what are surely vital tissues, and dolphins may use it as economically as possible.

That may be part of the reason Echo and Misha were reluctant to echolocate when we first started our experimental sessions—since they didn't need it in the tank, why wear themselves out? It certainly may explain how at first Echo managed to get around somewhat while blindfolded.

Neither Echo nor Misha seems to have done any full experimental trials in complete silence when blindfolded; they did, however, sometimes "shut off" their sound for periods as long as one or two seconds in a roughly eight-second run, while proceeding quite accurately toward the target object. On some occasions they even showed head-scanning behavior during such silent running. This seemed to occur most often on trials where the task was to retrieve a sinking ring; the ring may make enough noise as it moves down through the water column that the animal can simply find it on that basis.

I'm still looking at such questions as: When does the animal turn on his sound and for how long? How many click trains does he use per trial and how many clicks per train? When do pauses occur? What is the dolphin doing and how is he oriented during a pause? Artur Schnabel once said, "The notes I handle no better than many pianists. But the pauses between notes—ah, that is where the art resides!" I suspect something of the art of echolocation may also reside in the pauses.

I'm also working on comparisons between Echo and Misha on these tasks. This may give some hints as to what patterns are central to the echolocation process and what may be matters of individual style. For example, the Boys seemed to take approximately the same amount of time to complete a task, but they used the time differently. Echo, not surprisingly, would take off like a shot from the training station and then spend more time at the end braking and steering himself into position. Misha tended to take a more leisurely swim but then devoted less time to final maneuvers.

The popular literature often suggests that dolphins may be able to use their echolocation as a sort of language. Since their echolocation system gives them a detailed picture of the world around them, the argument goes, they might be able to project images directly to one another by recreating the sounds associated with particular objects or events. In theory, they might even be able to make up stories that way, or perhaps create novel, abstract images—acoustic artwork of sorts.

I presume that's what John Lilly had in mind when he referred to "Delphinese," the language of the dolphins. He considered this language to be analogous to human speech, though with "acoustic pictures" as the basic elements, rather than words of the sort we're familiar with.

I know of no evidence to support such a view. None of the "hard" scientific kind; none of the "soft," subjective, anecdotal kind, either. By that I mean I haven't encountered any situation or interaction thus far that has made me think, "Yes! the best possible explanation for what just happened is that dolphins must have a language comparable to ours." I do give credence to such "squishy" evidence—the sort of observation that isn't scientific in and of itself but that may nonetheless lead you to formulate questions that can be addressed scientifically.

I have looked for such evidence, believe me. I did so want to talk with the dolphins. The notion that they might be able to convey pictures directly to one another was especially appealing— they could avoid all the messiness, the ambiguities and imprecisions of human language.

The more I've learned of dolphins, of their communication and echolocation, however, the less I've come to believe in a "Delphinese" we'll one day be able to decipher if only we find the proper Rosetta Stone. I see no *Star Trek*-style instantaneous translation devices in the offing.

One reason I'm skeptical about dolphins' using echolocation as a language by exchanging acoustic pictures with one another has to do with the sort of movement patterns we looked at in our research with Echo and Misha. A dolphin might have a hard time conveying an accurate acoustic image of something he'd echolocated on if the other party didn't also have access to his proprioceptive informa- tion—his internal sense of his own movements.

That's been a long-standing problem in the study of vision: how do we manage to see the world as stable and unmoving when we ourselves are moving, our eyes are moving, and hence the images received on the retina (the sensory surface of the eye) are moving? We somehow have to subtract our own movements from the mov- ing image we receive to end up with a constant picture of the world. The problem is even more complicated with touch. If, for example, you're exploring some new object, you may run your fingers over it in all manner of ways, so your touch "image" of the object is anything but stable; again you have to subtract your movements. If, say, we developed some device that allowed me to feel the sensations you were getting in your fingertips as you ran your hands over an object, I think I'd end up with a largely incomprehensible flurry of sensory impressions; I'd have a hard time resolving them into a specific object if I didn't also know how your fingers had been moving.

I would expect the same with echolocation. To make sense out of the returning echoes, a dolphin also needs to know what was sent out and how it was sent. Our work with Echo and Misha indicates that an echolocating dolphin is moving his body in all manner of ways, which in turn moves the sound beam around. Since both the frequency and amplitude of the sound vary depending on how far they are from the center of the beam, another dolphin might have difficulty forming a clear image from the incoming sounds alone.

In this sense the interpretation of echolocation information may be context bound; the sound patterns are intimately connected with the movement patterns. Misha might be able to read a fair amount from Echo's signals, for example, if he's there swimming alongside him. But I'm not at all convinced that Echo could project a clear image of some object, much less anything more complex, simply by swimming up out of the blue and rattling off a series of click trains to Misha. The dolphin equivalent of sitting around the campfire swapping ghost stories via echolocation signals seems to me unlikely. For all their ambiguity and imprecision, that's the beauty of words—they can transcend the immediate context of the event being described.

Along with the question of whether dolphins have a language of their own comes the question of whether or to what extent they can learn human language. John Lilly was so impressed with their vocal mimicry of human speech that he worked for a while trying to teach them to talk. (That's more or less the basis of the movie *Day of the Dolphin,* in which a pair of dolphins learned to speak English.) That never panned out.

Lou Herman and his colleagues at the University of Hawaii have carried out a more successful exploration of language skills in dolphins. For more than a decade they have been studying the ability of a pair of bottlenose dolphins to understand sentences in one of two artificial "languages." One dolphin has learned an arm-gesture language, the other an acoustic language based on computer-generated whistles. Both dolphins have proved highly proficient at responding correctly to such commands as "surfboard fetch hoop," which in the acoustic language tells the dolphin Phoenix to take the surfboard to the hoop. "Hoop fetch surfboard," in contrast, means take the hoop to the surfboard. Both dolphins can respond appropriately to such reversible commands. Herman argues that this demonstrates their sensitivity to word order, to syntax.

Herman has been criticized by some for calling what his dolphins are doing "language" and for using such linguistic terminology as "sentences" and "syntax" to describe their behavior. I find his work fascinating and informative, revealing a great deal about dolphins' cognitive abilities, but I'm inclined to agree that it may be stretching things to call what they're doing "language." I'm not so much concerned with their syntactical abilities, however. My objection to calling it language is that the dolphins themselves don't have the opportunity to use it as such; they can demonstrate their comprehension by following commands, but thus far their "language" provides them no means of carrying on a conversation. Conversation, an interchange of some sort, is, I believe, at the heart of language.

In this regard I'm more impressed by the language abilities of various chimpanzees (especially the pygmy chimp, or bonobo), gorillas, and orangutans who have learned sign language or artificial

communication systems. Some of them very clearly use their ac-
quired language skills to carry on conversations with humans and,
in some cases, with each other. Dolphins may be able to do likewise, but they haven't yet had the opportunity. Chances are the
primates, as our closest relatives, have cognitive skills more similar
to ours, more akin to the ones that underlie our language, and
therefore are likely to do better with human-type languages than are
dolphins. Regardless of their relative degree of intelligence, dolphins' cognitive processes are likely to take a more alien form.

The answer to the question of whether dolphins have language
depends, in part, on how you define "language." We tend to use
the term in two ways: In technical, linguistic parlance it refers only
to what humans do, with words, sentences, and plenty of syntax; in
everyday speech, however, we use it more broadly, referring to body
language, the language of love, and so forth. My 1969 edition of
the *American Heritage Dictionary* lists among its definitions: "the
manner or means of communication between living creatures other
than man: *the language of dolphins.*" That definition is missing
from the 1992 edition, alas.

Linguists generally want to define language only in human
terms, to stake our claim as the sole language users. I prefer a
broader definition, one that includes such things as the waggle
dance of the honeybees and the "pep rallies" of wild dogs. (Before a
hunt they twitter and yowl together, with increasing excitement,
presumably "psyching" themselves up and assuring social cohesion
for the chase ahead.) Human language most certainly is unique; no
other animal on Earth, including dolphins, in my opinion, partakes
of our sort of language. I find it more useful to distinguish between
human language and other kinds of language than simply to say
there's language—which only humans have—and nonlanguage.

Human language surely is much broader and more flexible in its range and scope of subject matter than any other manner of language we know, but nearly all aspects of it may be found in some form and to some degree in other animals.

I can't believe that language is the "evolutionary break" some linguists would have it, a wholly human creation—sprung fully grown from the head of man alone, like Athena from the head of Zeus. It must be part of the same evolutionary tree as the communication systems of other animals. We may be out on a limb by ourselves, but surely human language draws life from common roots. We may be too quick to assume our own uncommonness and thereby lose something of our ability to "make common" with other species.

Dolphins possess a variety of means of communication, both vocal and nonvocal. We still don't fully understand the nature, scope, or significance of any of them. While I doubt any of them translates into "Delphinese," I do suspect that these multiple communication systems allow dolphins to communicate a great deal to one another, in a variety of contexts and at varying distances. As with us their communication may be multileveled, using several media at once. Given dolphins' highly social, highly cooperative nature, the complexity of their behavior, and the amount of behavior that depends upon learning, I'd be most surprised if their communication weren't more complex than that of your average animal.

Dolphins, sound-oriented creatures that they are, undoubtedly rely most heavily on vocal/acoustic channels for their communication. As a result of the streamlining of their bodies for rapid swimming, they lost much of the mobility of face, body, and appendages commonly involved in communication among terrestrial animals. Where land mammals may use facial expressions, hand gestures, wagging tails, flaring nostrils, and the like to get their message across, dolphins have to depend more on vocalizations. They have a wide array to choose from.

Dolphins have an extensive repertoire of sounds. Besides their whistles and echolocation clicks, they make a whole range of complex sounds—squawks, blats, "raspberries," and the like—that are collectively termed "burst-pulse sounds." These are perhaps the most familiar to our ears, since they are similar to human speech in both structure and perceptual quality. They also seem to convey a great deal of emotion and are particularly common in such contexts as aggressive encounters, sexual activity, or in response to novel or unexpected objects or events. Josephine has a particularly impressive "vocabulary" of burst-pulse sounds and can "cuss" with the best of them.

Dolphins can simultaneously produce pulsed sounds, which include both burst pulse and echolocation clicks, and whistles. Whistling and echolocating at the same time may be useful when two or more individuals are trying to coordinate their efforts for cooperative fishing, for example. Combining whistles and burst-pulse sounds could add another level or layer to the communication, a bit like talking and writing on the blackboard at the same time. Adding to this are their many percussive sounds, produced by slapping or splashing various parts of their bodies against the water in creative ways.

Dolphins do engage in "conversations" using their natural forms of communication. They often whistle in antiphonal fashion—when one whistles, another calls in response. They also engage in various sorts of "chorusing" behavior. Dawn Goley regularly observed such choruses among the Lags, both early in the morning, starting somewhat before sunrise, and again in the early evening. All the Lags would carry on, using a wide variety of burst-pulse sounds intermixed with whistles, and making quite a racket. Echo and Misha did the same sort of thing, getting particularly noisy and rowdy at dawn, though with just two of them the effect was somewhat less raucous. Ken described a similar phenomenon among spinner dolphins, which he called the "Yugoslavian news report," since it was like a running commentary in an unfamiliar language. This chorusing occurred just as the group was coming out of rest and preparing to start the next round of activity. The purpose seems to be to coordinate the group, to get everybody at the same level of wakefulness and awareness before proceeding. Dolphin schools often seem to function as leaderless societies, with no one to order, "Okay, everybody up and at 'em."

Dolphins also may be capable of a level of communication that goes somewhat beyond the immediate, here-and-now encounter generally considered the norm among nonhuman animals. My evidence in support of this contention is of the "squishy" kind mentioned earlier.

We were still working with Josephine at the time, trying to get her to wear eyecups and do echolocation runs. She tended to swim a very meandering path toward the target, whereas we wanted her to take a straighter course. We decided we would lift the target out of the water if she started wandering—that way she couldn't complete the task "incorrectly." Jo wandered. I lifted the target out of the water. Jo came over to where the target had been and glared up at me accusingly, as if to say, "Didn't you just ask me to touch that thing?" I felt as though I'd invited somebody to sit down and then pulled the chair out from under her.

Jo was furious. She swam off and headed toward the thirty-foot pool, where Arrow was temporarily housed while we worked with Jo, with only a net gate separating them. She let fly with the nastiest-sounding string of burst-pulse noises, directed into the tank where Arrow was, not back at us miscreant humans. Jo didn't seem to be taking out her anger on Arrow, however—a go-home-and-beat-the-wife response—but rather to be conveying to Arrow some sense of her anger at us. I doubt it would translate into words as such—"And then they pulled the frigging target out of the water . . ."—or even mental pictures of what happened. Rather, I think the message was one of emotion and of relationship. Anthropologist-philosopher Gregory Bateson argued that communication among preverbal mammals is largely about "the patterns and contingencies of relationship." My admittedly subjective sense was that Jo was letting Arrow know, in no uncertain terms, the current status of her relationship with us.

If I'm right, I think that indicates a higher level of communication than if she were communicating directly to Arrow about the status of their relationship. For example, if I walk into a room, bump my shin on a chair, and yell "ouch" (or something more emphatic), it's simply broadcast communication, not necessarily directed at anyone in particular. If I walk into a room, you kick me in the shin, and I start yelling at you, that's direct, one-to-one communication. But if when I walk into the room and you kick me, I then turn to another person and start complaining about what a mean-spirited so-and-so you are, that's something different; that's communication a step removed from the relationship I'm talking about. I suspect that level of communication would develop only among animals that have highly complex social systems—not just dolphins, but probably at least higher primates, perhaps elephants and some social carnivores as well. It doesn't require language as we know it, but it does represent a communication that goes beyond the immediate, direct interaction.

While I doubt that dolphins use their echolocation in a language equivalent to ours, I do believe it is important in their communication. Ken describes echolocation as "autocommunication," communication with the self—or with the environment. The dolphin emits a sequence of sounds that are modified by the environment and returned to the animal as a message containing information about the environment.

There are also at least a couple of ways in which echolocation could function in interdolphin communication without forming any sort of language. First, they may well "eavesdrop" on each other, listening in on the next guy's echolocation. Ken found that among spinner dolphins the animals appear to take turns serving echolocation duty. Another individual not on duty at the time probably could keep tabs on what's out there by listening to his buddy's click trains. He may not get the complete picture his companion does, but by listening to the rate, and especially changes in the rate, he could at least tell if there's something potentially interesting or threatening out there.

Second, there's speculation that dolphins could use their echolocation in a fashion similar to modern ultrasound machines. Dolphins thus might be able to look inside each other's bodies, reading their schoolmates' health status, perhaps even their emotional states. Michelle said that when one of the female Tursiops at Brookfield was pregnant, a young calf spent a good while inspecting her belly. I've also heard anecdotal reports of dolphins in swim programs intensely scanning pregnant women.

Ken is doubtful. While he believes a dolphin might well be able to detect the motion of a fetus inside another dolphin, looking into the body and inspecting each other's internal workings may be another matter. In one study he tried projecting sounds through the blubber coat of a dolphin and into the muscle tissue beyond. It didn't work. The blubber and muscle aren't completely attached to one another, so the two layers can slide with respect to each other, forming an acoustic barrier almost no sound can pass through.

On the other hand, ultrasound machines do manage to pass sound through the blubber and get images of the dolphins' internal physiology. Such machines employ even higher frequencies than dolphin echolocation, however—measured in megahertz (millions of cycles per second) as opposed to kilohertz (thousands of cycles per second)—which could make a difference. The notion isn't so far-fetched, though, and many reputable scientists consider it a definite possibility.

I also believe there are some interesting parallels between dolphin echolocation and human language. Dolphin echolocation is a highly complex and sophisticated use of sound; in some respects it may rival, even surpass, human speech in this regard. It's a central feature of the dolphin mental landscape, which I suspect may form the foundation or organizational base for dolphins' lives—their thoughts, their memories, their communication—as language does for us.

Echolocation isn't simply an innate skill but involves considerable learning by youngsters, just as language does for us. Calves must learn it at their mother's flipper, so to speak, practicing their clicks and discovering how to interpret the information they receive back. At first they sputter and slur their clicks like a babbling child.

Furthermore, dolphins seem to have social rules regarding echolocation, or "echolocation etiquette," as Ken calls it. He found that among captive spinner dolphins, when an echolocating dolphin approaches one of his companions, he'll shut down his echolocation gear, or at least turn the volume way down; then once he's past his buddy, he turns the sound back up again. This may be a matter of protecting each other's hearing, so crucial to a dolphin, by not noising off too close by. On the other hand, dolphins are known to blat right in each other's faces when angry. Maybe it's simply impolite to scan somebody else's innards. "It's not nice to stare. . . ."

Finally, I believe the sort of collateral behaviors we recorded from Echo and Misha are a vital component of echolocation, that the rhythms imposed by its movement patterns may provide the animal with a means of structuring the sound stream, of dividing it into analyzable segments, in a manner somewhat analogous to the way we use the rhythms of our breath stream to impose patterns of stress and intonation on our speech. In echolocation, as in human speech and music, there may be much truth to the old song, "It don't mean a thing if it ain't got that swing."

People are often disappointed when I tell them I don't think dolphins have language as we know it. I guess I was, too, at first. I've been learning more and more to appreciate them for what they are, though, without being disappointed that they're not what we are. I remember reading a science fiction story once in which humans felt sorry for some alien species because they lacked eyesight. It turned out, however, that they had some other sense we couldn't begin to conceive. One of the humans got the opportunity to experience the sense, but only temporarily; he was so amazed by it that he felt utterly bereft thereafter without it.

We humans tend to regard our language that way, too. I've always felt we are perhaps a bit too proud of our language, too quick to assume our superiority over the "dumb beasts" around us—as though they envied us our speech and would choose it, if only they could. We may consider them deprived of language, but we might do well to remember that there are many ways of sensing the world; we may be comparably "deprived."

9

Homeward Bound

Ken had forewarned us in December, just as Echo and Misha were completing their echolocation runs, that the Boys' departure might be imminent. Our plan from the outset had been to return them home within two or three years of their capture. But by the end of 1989 serious financial strain forced us to consider an even earlier release. Dolphins are an expensive venture—maintaining the facility, paying (however minimally) at least some staff members, keeping the dolphins in fish, monitoring their health, and, if necessary, treating any ailments. At one point Randy estimated the monthly cost at three thousand dollars just for salaries, food, and vitamins, not including either general facility maintenance or health care.

Our original grant for the project, from the Office of Naval Research, had presupposed that we could complete the project with Josephine and Arrow, the two dolphins the Navy had given us. It hadn't taken into account the considerable costs of transporting Jo and Arrow to Florida, catching Echo and Misha, and bringing them back to Santa Cruz, much less the expenses involved in returning them home again afterward and conducting an extensive follow-up study. Nor had we factored in the additional time, and therefore additional maintenance expenses; our two years of work with Jo and Arrow yielded minimal results, and we had to start all over again with completely naive animals, beginning with the very basics of training protocol. Funding for such projects was scarce to begin with, and Ken, now on the verge of retirement, was in no position to apply for another grant.

We couldn't afford to keep the dolphins much longer. We couldn't afford to get them back home again, either. Flying Tigers, the outfit that had originally flown Echo and Misha from Florida to California, had since gone out of business, bought out by Federal Express. Federal Express told us, "We don't do animals." Charter flights cost about twenty-eight thousand dollars, which would eat up most of our remaining funds, leaving very little for the follow-up study, so vital to the whole sabbatical project.

We got any number of leads on free or inexpensive flights, but one after another they fell through. Ken first sought to arrange for Echo and Misha, plus people and gear, to hitch a ride on a Sea World plane that would be transporting some of their dolphins from San Diego to Orlando. The prospect was less than ideal on several counts. For one thing, it would mean first flying or trucking the Boys to San Diego and then finding temporary accommodations for them there and possibly in Orlando as well—not an easy matter. It also meant a mid-February departure date, which was certainly rushing things on our end. Worse yet, we really didn't want to release the dolphins that early in the season—we'd still be facing winter storms and fog, which could prevent the follow-up crew from going out in search of the dolphins. We'd prefer to wait for spring or summer to improve our chances of tracking and spotting the animals regularly following their release. Ultimately, the question was moot; we simply couldn't get all the necessary permits in time for the Sea World flight.

Ken got an offer from a fellow who restores amphibious aircraft, but the flight would still be expensive and the plane too noisy and too slow for Echo and Misha's comfort. We had thought the Navy might fly the Boys home, since they transport their dolphins regularly and our echolocation research was a Navy-sponsored project (though the release study was not). No go. Our hopes for a company that ships horses around the country, and for an organization in Georgia that has conducted aerial surveys of right whales, were similarly dashed.

We also checked with various commercial airlines to see if they'd be willing to fly us free, making use of it for their promotional purposes, of course—fly the airline that freed the dolphins. But most airlines don't allow people in the cargo hold, and we couldn't transport the dolphins without people there to attend them continuously.

At least for the moment the charter flight, expensive as it was, seemed our only option. We went ahead and started making arrangements to fly the Boys home at the end of March. We'd just have to figure out some way to come up with money for the follow-up study.

We all tried our hands at fund-raising—just about everything short of door-to-door soliciting. We sold T-shirts and sweatshirts. One of the dolphinettes offered her artistic skills and created the design—a picture showing Misha leaping and Echo with his backgammon-board forehead, with the words "Release 1990" above and "Misha & Echo, Homeward Bound" below. Holly arranged for the printing. We all sold them to our friends and families and anyone who would listen. The Long Lab gift shop sold them. We posted a display in a local restaurant chain. We also started work on a "Send the Boys Home" promotion loosely based on the adopt-a-whale programs—anyone donating a certain amount would get a newsletter about the release and Echo and Misha's progress thereafter, plus a photograph of them in the wild. Jill did most of the planning and organizing; I wrote the brochure. The scheme never got off the ground, however, at least in part because the university didn't approve.

The UCSC bureaucracy, by and large, was of little help in our fund-raising efforts; if anything, it was more of a hindrance. The university didn't hesitate to display Echo and Misha's pictures prominently in catalogs and promotional materials, and to bring potential big donors out to the lab for an audience with the dolphins. They didn't see fit, however, to help send them home. I guess that makes sense—development offices are in the business of acquiring money to develop projects, not to dismantle them. They didn't want us competing for funds with other programs, nor did they want us soliciting from their major contributors.

Besides worrying about transportation for Echo and Misha, we were also concerned about their transition. We had no intention of flying them to Florida and just dumping them back into Tampa Bay. We had to work on preparing them for release in various ways, first at Long Lab and then at a "halfway house" in Florida. After transport but before release we would keep them in a bay pen, similar to the pen we constructed at Ruskin at capture time, for at least a week or two. This would give them a chance to reacclimate to their home waters—all the sights and sounds and tastes of Tampa Bay—before venturing out on their own. It would allow us to make sure they handled the transport well and were in good health. We could also determine that they were able to catch and eat live local fish before we sent them off to fend for themselves. Ideally, the pen would be back at the wharf at the Bahia Beach—Echo and Misha's last point of reference in Tampa Bay before their sojourn to California. That way, when the time came, we could just open the pen and let the Boys swim away if they chose, or stick around awhile if they preferred. They could even come back to be fed for the first couple of days if need be.

This plan led to a barrage of permit hassles, however. Randy had spent much of the fall of 1989 and early winter of 1990 filling out applications for the various and sundry permits we would need. We needed permits, it seemed, from anybody and everybody, every step of the way. The permit for capture did not cover release. The basic federal permit for the release came from the National Marine Fisheries Service (NMFS), which oversees virtually all dolphin business nationally. We also obtained a permit from the Florida Department of Natural Resources (DNR) for transporting dolphins within the state of Florida. The biggest hassle, though, involved permits to build the temporary holding pen. Apparently the permit process for construction at the water's edge is geared for more permanent structures—wharves and seawalls and the like. We just wanted an inexpensive stopgap structure, one that therefore wasn't designed to meet all the building codes. Half a dozen state and local regulatory agencies required application forms, diagrams of the proposed structure, and explanations of what we were up to. Ultimately, all the interested parties gave us the permission we needed, or exemptions from it, though the process took several months.

One important requirement for release was that Echo and Misha must be able to catch and eat live fish. You'd think that would be easy. No way.

We started reintroducing the Boys to live fish in January, occasionally putting small schools of frigate mackerel into the tank with them. They seemed downright offended. Neither Echo nor Misha would even go near the critters; far from pursuing the fish, our dolphin duo fled from the intruders, swimming to the opposite side of the pool and hovering at a safe distance—as if the mackerel were the predators and they the prey. When we tried hand-feeding them fish that were stunned but still living, both Echo and Misha simply spat the things out, like some kid gagging on his broccoli, and refused to continue their training sessions.

We also periodically presented them with frozen (thawed) Florida mullet to remind them of the food back home. They were unimpressed. No live mackerel, and no dead mullet, would pass their lips.

The Boys appeared to be very conservative in their eating habits. They had reacted similarly when we first presented them with cut fish. Jo and Arrow as well would often quit eating temporarily when we changed their food in any way, giving them smelt instead of capelin, for example, or sometimes simply a new shipment of their usual type of fish. Undoubtedly in the wild it's vital for them to know what they can and can't eat. It may have social ramifications as well. "Maybe it's like the Japanese tea ceremony," Holly suggested—"you don't just pour a cup of tea. They almost seem to have traditions, which we don't understand, around the food they eat. If you break the tradition, you're committing a cultural faux pas—it's just plain rude."

We never did get Echo and Misha to accept live fish at Long Lab. We finally gave up trying and decided that would be one of the main goals during their "halfway house" stay before release. Back in Florida we could give them live fish of the variety they normally would eat, and that might ease matters some. If they still refused live fish, we had a serious problem. We simply couldn't release them until they demonstrated convincingly that they could pursue, catch, and eat their own meals in the wild. But then, that's part of what this study was about—can dolphins who have lived in captivity learn to be wild animals again?

Meanwhile, we'd continued with the research training, even as we prepared for the Boys' return home. Although we'd recorded all the basic echolocation runs we needed, we still hoped to conduct a test of the "beam-steering" hypothesis. Several lines of evidence suggest that dolphins may be able to move their sound beam internally, within their foreheads, at a rate up to five times as fast as they could move their heads in a scanning motion. To test this notion, we wanted to keep the dolphin stationary in front of an array of hydrophones as he performed an echolocation task. If we found evidence of the sound beam moving across the hydrophone array (higher frequency and amplitude at one hydrophone and then at another) while the dolphin's head remained motionless, it should mean the dolphin was moving the sound beam from inside his head—internal beam steering.

To get the dolphins to sit still for all this, Ken built another of his famous contraptions, which was supposed to allow the animal to rest comfortably in one spot as he echolocated. The device had what looked like a large set of handlebars that the dolphin could rest his belly on and a smaller set of handlebars in front to rest his chin on. It had a sort of kickstand that came out underneath (if all went well when we tossed it into the tank) with a foot plate that sat on the bottom of the tank, anchoring the whole thing; the handlebars stood about four feet below the surface. Officially, in the training records, we referred to it as the "beam-steering apparatus." Unofficially, we dubbed the thing "the bicycle from hell." I suspect Echo and Misha had an even worse name for it. They wanted nothing to do with this newest monster. "Rest on *that*? You've got to be kidding." As with just about anything new, their first response was to flee. They didn't seem to grow much fonder of it over time, either.

Toward the end of February, Misha started dropping his fish, refusing to eat his usual amounts. Holly was concerned that the live-fish attempts and the "bicycle from hell" might be too stressful for him. She was convinced we needed to stop the training. I agreed, though I was reluctant to forgo the beam-steering study, which Ken had all along considered the newest, most exciting aspect of our project. We arranged a meeting with Ken in early March to discuss the matter. Knowing that for all his folksiness Ken can also be quite formidable at times, Holly had prepared a list of all her reasons, twelve points in support of her view. Ken, in his formidable mode that day, was not persuaded. He wanted to continue the research. I folded in the face of his disfavor. Holly got up and left the room, in fury and in tears.

Ken seemed genuinely puzzled by her reaction. I left to go after her. "I thought you were with me," she accused as I approached. "I thought you agreed." She was right. I went back to talk to Ken, this time telling him simply that I thought we should stop the research. He conceded. No more beam-steering study. No more bicycle from hell. Just work on getting the Boys ready to go home. The welfare of the animals always came first to Ken; he'd just needed more convincing, apparently, that Misha's health was indeed at issue.

Misha resumed eating and acting like his normal self. We finally had all our permits in hand and were set to send the Boys home at the end of March. Jay Sweeney came up from San Diego two weeks beforehand to give them a preflight physical and certify them fit for transport. During the exam we freeze-branded both Echo and Misha using liquid nitrogen, a painless procedure similar to the one doctors use to freeze off warts. We gave each of them a three-inch-high identifying number, one on each side, just below the dorsal fin. This would ensure that the follow-up crew could identify them unequivocally out in the wild, where all you normally get is a quick glimpse of an animal's back. Misha received the number 202, Echo 204.

Randy regularly freeze-brands the dolphins in his Sarasota study group as part of the capture-release program so the animals can be reliably identified from the boat whenever they are sighted. Males are given even numbers, females odd, and they are generally numbered in order of capture. By this scheme we numbered Echo and Misha backward—Echo had been caught first and so should have the lower number. Misha is Randy's favorite, however, so he tends to put Misha first in all things. He and some others consistently say "Misha and Echo," whereas others of us say "Echo and Misha." I do the latter, though that isn't meant to favor Echo; it just makes sense alphabetically. Besides, Echo was not only first caught but first named.

Jay also pulled a tooth from each dolphin, under a local anesthetic. This allowed us to determine their ages more definitely. Dolphins' teeth, when cut in cross section, show growth rings, something like the rings on trees, that correspond to their age. Their tooth rings confirmed our earlier estimates. Misha was then nine, Echo eight—so they had indeed been seven and six, respectively, when we caught them. Randy's team also pulls a tooth from any animal in his Sarasota study group for whom he doesn't know the year of birth from direct observation. This age information is vital for making sense out of the population's social and behavioral patterns.

Jill was slated to be the primary researcher on the follow-up study. She was planning to get her master's degree in marine science at UCSC, with Randy as her adviser, and to use the Echo and Misha follow-up work as her thesis study. She had just quit her job as a waitress and had given up her apartment in preparation for driving cross-country to Florida to be there when the dolphins arrived. She was set to leave the next day when Randy called from Chicago telling her to hold off. A possible epidemic threatened dolphins along the coast of the Gulf of Mexico: an unusually large number had been found dead—several hundred since January—all the way from Texas down into Florida. The death rates weren't as high in the Tampa Bay area, at least so far, but we didn't want to take any chances. The National Marine Fisheries Service advised us to wait until they could learn something about the cause of the deaths before attempting to release Echo and Misha.

That was the last straw for Jill. She'd already had more than her share of hassles with the university over her status in the marine sciences program, which was still unclear. She'd suffered through the various vicissitudes of the project, all the waiting and all the uncertainties. Now she was jobless, homeless, and had no project to go to for who knew how long. Randy urged her to go ahead and move to Florida and wait it out. He had managed to arrange funding for her to carry out the follow-up study. But that wouldn't cover her during the waiting period, which was as yet indefinite. She lacked resources to fall back on—she'd just been a starving student working her way through school as a waitress.

Jill decided she'd had enough. She quit. Angrily. Unhappily. She left the crew and left the lab. Jill never saw Echo and Misha again.

This left us without a follow-up person. We really didn't want to start with someone brand new. We needed someone who already knew the Boys well, knew their behavior and their "dolphinalities" as well as their physical appearance. For a while the plan was to have several of our crew members trade off, a couple months at a time; no one could afford to take off and devote the whole year to it, as Jill had intended. This was less than ideal, however, since we'd lose something in consistency. We could only hope to find a volunteer in Florida who'd work for the duration of the project, providing the necessary constancy.

Ultimately, we were fortunate enough to get Kim Bassos, one of the volunteers who had joined the crew when the Lags first arrived, to take on the project. She would be graduating in June and had just been admitted to the master's program in marine science, hoping to study dolphins under Randy's direction. A statuesque beauty with a keen mind and a gentle spirit, Kim was described by one friend as "a thinking man's Daryl Hannah."

The Gulf-coast dolphins weren't the only ones having problems. In late March, about the time we'd been hoping to send the Boys home, Misha began eating irregularly and acting lethargic. On April 1 we conducted a medical exam, taking blood, fecal, and stomach-fluid samples. After consulting with Jay Sweeney, Dave Casper prescribed a course of medication, including an antibiotic and a drug designed to control gastric-acid secretions.

Within the week Misha was eating much better, but then his appetite dropped off again a few days later. Given his symptoms, the lab results, and the cyclic, up-and-down course of the disease, Jay figured the most probable diagnosis was an ulcer in the region just past Misha's second stomach. (Dolphins have multiple stomachs, like their ungulate cousins, the exact number depending on species and on who's doing the counting.) His appetite loss back in February may have been part of this same cycle.

A subsequent ultrasound exam revealed an unknown mass in the back portion of the stomach, which could have been a granuloma, an aggregation of inflamed tissue often associated with such an ulcer. Given the location, however, the mass could also have represented an infection of the pancreas, a swollen lymph node, or even a tumor; we had no way of knowing for certain.

An ulcer was by far the most likely possibility. Given the nature of their digestive tracts and because they digest whole fish, complete with bones and scales, dolphins are particularly prone to ulcers. Along with respiratory ailments, that's one of their major weak points with respect to health. Whereas dogs and cats, for example, almost never get ulcers, humans and dolphins quite often do. This is just as true of wild dolphins as of captive ones; necropsies performed on many animals found dead on the beach show clear signs of ulcers.

By the end of April, Misha seemed much better. His appetite

improved. His blood values returned to normal. Meanwhile, the death rate among the Gulf dolphins had started to subside. We even had a lead on a free flight through Delta airlines. Things were looking up. We started making plans for transporting the Boys back to Florida in midsummer, providing Misha was healthy and off all medications for at least a month before the transport date. We might even be able to coordinate the timing of the release with Kim Urian's wedding, which was to take place in Florida in July. She was marrying Andy Read, a cetacean biologist who often worked with Randy on the Sarasota project and who had helped with Echo and Misha's capture. Randy, who belongs to the same mail-order ministry as Ken, would perform the ceremony.

Then in the latter part of May, Holly saw something in Misha's eyes—or maybe didn't see something—that roused her suspicions. Eyes may be windows not only to someone's soul but also to his health. Usually, decreased appetite is the first sign of illness in a dolphin. Additional signs we always looked for included a decline in energy level, decreased willingness to participate in sessions, and reduced interactions with people or each other. But the eyes were also a clue, especially with Misha, it seemed. His eyes would get squinty, or they'd get dull. Or both.

His appetite was good. His behavior seemed normal. But Holly had a gut feeling that something was wrong, like the sixth sense a mother may have about her children. Dave Casper came in on May 21 and asked, "How do you think Misha's doing?" For reasons she couldn't explain ("I hate this sort of stuff," she says, "because the 'dolphin huggers' love it"), Holly told Dave, "I think we should take blood, and I think we should do it today." Sure enough, his blood results were far worse than his behavior had indicated.

The behavior soon caught up, as he again started dropping his fish and acting lethargic. Then he quit eating altogether. If he wasn't eating, we couldn't get his medication into him in the usual fashion—pills stuffed into the gill slits of the fish. That meant we had to catch Misha regularly and give him fluids and medication through a tube placed down his throat.

He lost weight; you could see hollowed-out areas below his dorsal fin. He no longer had that wonderful, smooth, taut dolphin feel to his skin, either. Holly described him as feeling like "Saran Wrap over cottage cheese."

It was a grueling time for everybody concerned. For a while we were tubing him three or four times a day. That meant clammy wet suits and cold water for us, and restraining an unhappy dolphin. For Echo it meant suffering the indignity of finding himself beached, lying on the bare tank bottom several times a day. Generally we had a couple of people hold up one of the net gates to keep Echo at bay while the rest of us worked on Misha. Misha, of course, got the worst of it—not just disappearing water but people holding him, a tube down his throat, and, for a time, a shot below the dorsal fin twice a day. After a while we got the whole business down to an art. We didn't even have to lower the water all the way, but could manage in waist-deep water. That made it easier on the dolphins—more freedom—but also gave them more room to thrash around, increasing the danger for us.

By the end of May, Misha started objecting to the injections and resisting more when we caught him for tubing—a sign he was feeling better, perhaps, or simply that he was getting very tired of this whole business. Over the next week he started eating better, and as of June 5 we were able to dispense with the tubings. His behavior returned essentially to normal, except that we hadn't seen any sexual interaction between him and Echo since March.

We continued his medication through June and July and gave him weekly medical exams, taking blood, fecal, and gastric samples each time to monitor his progress and make sure there was no relapse. An ultrasound exam in mid-July showed the abdominal mass—whatever exactly it was—to be much smaller and less well defined.

Misha had lost forty pounds between the mid-March exam and

the one in mid-July, dropping from roughly 344 to 303 pounds. About half that was probably normal summer weight loss, however, since Echo dropped twentysome pounds over the same period. By late August, Misha had regained nearly twenty of his lost pounds.

The dedication of everyone involved during Misha's illness was impressive. The crew, especially Holly, put in incredibly long hours. Dave Casper gave most generously of his time, and Jay Sweeney consulted with us regularly by phone as well as making the trip up from San Diego several times. Randy, though he was in Chicago, got blow-by-blow reports of all that was happening. He had perhaps the hardest job of all: making decisions and arrangements regarding what to do about releasing the Boys in the face of Misha's illness.

Although Misha never appeared to be in any immediate danger of dying, he was clearly a very sick animal. We weren't sure we'd be able to release him. We wouldn't do so if his condition continued, and given the roller-coaster course of the disease, it would be hard to know for sure if he were cured or just on an upswing—to be followed by yet another downswing. If he wasn't releasable, what would happen to him? And what about Echo?

Brookfield Zoo agreed to house both Boys temporarily if need be, and to keep Misha on a permanent basis if he turned out to be unreleasable. We would go ahead and release Echo as planned, with or without Misha. That wasn't the sabbatical we had in mind, but it would still be better than nothing. We'd lose the presumed advantage of releasing a natural pair; but then if this was to be a test case for future releases, it might be all the more realistic with just one animal. And Echo, at least, would go home.

Before Misha got sick, when we still thought the Boys would be going home in spring or early summer, Holly had made plans to move to Hawaii in early July. She would join two of Ken's graduate students who were studying wild spinner dolphins there. When Misha had his relapse and the Boys' transport was postponed indefinitely, Holly decided to stick with her plans anyway. She truly wanted the opportunity to do fieldwork. Besides, I suspect she'd had her fill of the whole dolphin-training business at that point. In many respects she got the worst of the deal as head trainer—the final crunch to get the research data, plus Misha's illness. She'd never been quite the dolphin enthusiast some of us were. "It's just not the whipped cream on my cake," as she put it. She was interested in getting the training experience, yes, but ultimately she'd rather work with dogs. She never felt entirely comfortable with dolphins, especially the Tursiops, and was frustrated by how difficult it is to understand them. She was more than ready to take her leave.

The trouble was, that left me as head trainer. I was terrified. I'm not a particularly good trainer. Nor am I much good at being "in charge," at being anybody's boss in any way. (That was part of the appeal of being a writer—I don't want to have a boss, but I don't want to be one, either.) Fortunately, I had Lisa Takaki, a truly magnificient trainer, to turn to. I also had a great crew, and especially Chris Ames as my primary cotrainer; though still new at training, Chris clearly was a natural at it. As it turned out, I'm extremely grateful to have had that opportunity to work with Echo and Misha so intensely at the end, to get to know them even better before they left. And if Holly got the worst, most difficult, period to work with them, I think I got the best. It was a marvelous summer. Misha was well on his way to healing—eating well, no more tubings, no more shots. The research training was over. Mostly, we all just had a good time.

We did do some serious work on various husbandry behaviors that would better allow us to monitor Echo's and Misha's health once they were in the pen in Florida. Specifically, we wanted them to present their flukes voluntarily and hold still so we could draw blood from them, to open their mouths and allow us to put a tube down their throats and obtain a sample of their stomach contents, and to give us a forced exhalation into a petri dish so that we could get a blowhole sample.

Chris worked predominantly with Echo; I focused on Misha. Chris, a freckled strawberry blond with a spunky, outspoken temperament, seemed a good match for Echo. (Her nickname, Turtle, was short for snapping turtle, a notoriously obstinate breed.) They seemed to understand each other, and she didn't put up with any guff from him. I felt more comfortable with gentle, soft-spoken Misha.

Chris focused on taking blood samples from Echo. She began with fluke presentation, first getting him to place his flukes on her lap as she knelt on the edge of the platform, and gradually increasing the length of time he would hold steady in that position— twenty seconds, forty seconds, a minute, and more. Then she brought in a second person, usually Howard or Lisa, to rub his flukes and start introducing props, the various bits of equipment involved or a facsimile thereof. First someone rubbed his flukes with a cotton ball and alcohol; next the person would poke at the spot to be injected, then snap it with her fingers and later with a rubber band—all approximations of sticking him with the actual needle. Echo never flinched. The poking and prodding didn't seem to bother him; the tougher part was simply getting him to hold still in the proper position. Then she brought out the syringe, sans needle, and the catheter that would be used in drawing blood. Finally, after a month or so of practice, Echo was ready for the needle itself. Chris held him; Lisa drew blood. Echo actually sat still for it—some of the time.

I worked with Misha, concentrating on the stomach tube. We did a similar gradual buildup. First I'd just conduct a regular session but have the stomach tube draped around my neck so he'd get used to seeing it. Then I started introducing it to him, showing it to him, touching it to his rostrum, touching it inside his mouth on mouth inspection. I also worked on moving my fingers farther back in his throat during mouth inspection so he'd get used to the feeling. Finally we tried the actual tubing—just a little way down his throat at first, and then farther, farther. Finally, I did manage to get stomach samples from him. He seemed to tire of it all after a while, though, and started clamping his mouth shut, refusing to open for the tube. We had to backtrack some, repeating earlier steps. He did accept it again, though he remained moody about it, sometimes allowing it, sometimes not.

Only a somewhat small proportion of our time was devoted to such less-favored behaviors, however. We spent most of our time on activities that were more fun for us and, it appeared, for the dolphins. We had them jumping hurdles, doing high jumps, swimming laps around the pool—dolphin gym class. Even this had a serious intent, though, and was part of the plan for getting Echo and Misha ready to go home: we wanted to get the Boys back into shape, since they had been leading a somewhat sedentary existence compared to wild dolphins. We'd started most of the exercises back in January, but they'd fallen by the wayside during Misha's illness.

We set up a hurdle, made of thin PVC pipe strung on a rope, across the middle of the large pool. For starters we had the hurdle well under water so the Boys could easily just swim over it; because they tended to freak out at anything new, we didn't want to make it too scary for them. Then we started raising it closer to the surface. Surprisingly, at first Misha was the bold one, Echo the scaredy-cat, as they had been back when they'd first arrived. Misha swam right over, no problem. Echo balked and turned back. Repeatedly. Finally, they got used to it and we started raising the hurdle above the water, just an inch or two at first and then up to a couple feet or so. Once Echo got the hang of it, he turned into the hurdle king. He'd sail over it handily. Misha started to balk. Even more often he'd swim under the hurdle on his way out from the training station and then jump over it on the way back. A "left-handed" dolphin, maybe.

They seemed to enjoy it, so we started leaving the hurdle set up between sessions to allow them to play with it on their own. Echo went over it again and again, back and forth. After a while, though, we noticed that he wasn't clearing the hurdle as neatly as he had been, but that his peduncle hit the hurdle as he went over. It dawned on us that he wasn't just being sloppy or losing steam but rather was doing it deliberately, rubbing his genitals on the hurdle bar on each pass. He'd turned the hurdle into a sex toy. We decided to put a stop to that. They had plenty of toys they could use in any fashion they wanted between sessions, and they had a rub line (a thick rope stretched across the tank), which they could use to scratch themselves, rub off their sloughy skin, or use in any kinky manner they desired. During sessions, though, we wanted them doing full, energetic hurdles. We started taking the hurdle out between sessions and reinforcing the Boys only if they cleared the bar without touching it. Back to nice, clean jumps in no time.

We also set up a "high touch," a ball suspended by a rope from a beam above the tank, which we would raise or lower to various heights. Here their task was to swim fast across the tank, leap out of the water, and touch the ball with their rostrum—a kind of behavior often seen in dolphin shows. They worked up to a height of just over eight feet above the water (in bigger, deeper tanks dolphins can jump a good deal higher than that), before Randy asked us to stop the high jumps. He was afraid the Boys could injure themselves in the process, and that was the last thing we wanted at this point. He also asked us to stop tailwalks. We were working on having them tailwalk for distance (Echo was up to three quarters the length of the fifty-foot pool) and height (as much of the upper body out of the water as possible). This isn't exactly a natural dolphin behavior, however, and Randy feared they might strain a muscle.

We continued with the hurdles, regular leaps, and various

sprints around the pool. We did "point follow" in which the dolphin was supposed to follow where the person pointed—so we'd run as fast as we could around the pool, dolphin following alongside. We also did "A to B," in which the dolphin was to station with his rostrum on a target held by trainer A and then, on signal, swim fast to a target put in the water by trainer B. When they started cuing in to where the other trainer was standing even before they were sent, we tried to keep them guessing by having the other trainer duck down below poolside and then pop up anywhere around the pool and toss the target into the water, so the dolphin had to search and find. Or we had a bunch of people around the pool, any one of whom might have the target on a given run. Emphasis was on speed in these games.

We also kept them jumping, working on height and trying to get them to do two or three in a row before we'd bridge (blow the whistle). And we started to intersperse brief "minisessions" between their longer feeds throughout the day. One of us would run out with a few fish, send Echo and Misha on a few leaps or whatever, and that was it for the moment. Then maybe someone else would have them do something else a few minutes later. We wanted to keep them on their flukes, as it were.

By and large, it was a playful, exuberant time for all involved. While they still had their off moments, comments in Echo and Misha's training books from that summer consistently reported things like "energetic and enthusiastic," or "great session." Misha seemed truly recovered and in high spirits.

Even gating was easy. We were gating them almost every day at this point, so the Boys would have a good part of the day in the large tank to get their exercise, and the Lags got it at night. Partly because of the regularity of it, but also I think because they were in good moods, gating went amazingly smoothly. We never needed the net, and usually the Boys went through the opening as soon as we lifted the gates out of the way.

We started thinking release again. The free Delta flight for Echo and Misha hadn't panned out. Somehow we weren't surprised. Fortunately, though, a woman who had handled the animal transports for Flying Tigers and stayed on with Federal Express went to work on our behalf. She was able to convince the company that, yes, they could do animals. They agreed to try their hand at transporting two dolphins in need of a ride home, and at a cost far less than that of a charter flight.

We still needed funding for the follow-up study, however. Misha's illness and the Boys' extended tenure at LML had further depleted our already scant funds. For a while Misha's medication alone had run fifty dollars a day. (Apparently pharmacists require a surname for prescription labels. The local Long's Drug, where we bought the medicine as well as other supplies, listed the Boys as "Misha Dolphin" and "Echo Norris.") Ken ended up having to sell off much of the equipment he'd accumulated over the years for his dolphin studies—including the project's pickup truck, the twenty-five-foot boat (the *Nai'a*) Dawn had used for her field studies of Lags, and various pieces of equipment from the bioacoustics lab—just to keep the dolphins fed and cared for during the remainder of their stay with us.

The International Wildlife Coalition donated three thousand dollars toward the release. Our major benefactor on that score, though, was Mote Marine Lab in Sarasota, Florida. They provided not only substantial monetary support, but boats and people as well.

We also arranged to set up our "dolphin halfway house" at Mote Lab. Given Misha's recent illness, we wanted a longer adjustment period before release—a month or so rather than one to two weeks as originally planned—so we could be sure they weathered the transport well and were completely healthy. The Mote site would be better protected from wave action, boats, and the public than the Ruskin site and gave us better access to veterinary care as well. Plus, the Mote staff would be catching the live fish Echo and Misha needed to relearn how to catch and eat. The only drawback was that come release time, we couldn't just open the pen and let the Boys swim away; we'd have to take them two hours by boat back to an appropriate location in Tampa Bay.

Brookfield Zoo generously agreed to pay Randy's and Michelle's salaries for the time they would spend in Florida directing the transition period and follow-up study. They also allowed us to use two of their dolphin transporters—the wet-transport variety Jay Sweeney had designed. They even went so far as to have their fiberglassing crew modify the transporters to fit Echo and Misha, according to measurements we'd taken during a physical exam, and to ship the transporters to Long Lab.

Joel and I had long been planning a trip to Italy (he'd accumulated enough frequent-flyer miles for two first-class tickets to Europe). We had originally intended to go in the spring, before the summer crowds hit. I begged for a delay, since at that point we thought Echo and Misha would be leaving, and I wanted to see them off and to be in Florida for their actual release. We decided to wait until after the summer crowds—the dolphins should be long gone then. We finally went ahead and ordered our nonexchangeable tickets. Departure date: September 11.

Misha's illness had postponed everything, of course. He seemed completely recovered by late July, though, and we ended his medication as of August 1. Randy scheduled the flight with Federal Express. Departure date: September 11.

At least the Boys' plane left several hours earlier than ours, so I could see them off, then dash home to get ready to take off myself. I wouldn't have been going on the transport anyway, and I couldn't

afford to go to Florida for the whole adjustment period. I'd be back
in plenty of time to go to Florida for the release, as I'd planned all
along.

When we started the Boys on their exercise regime and shifted to
daily gatings, we also decided to forgo the watertight gates between
pools. We still kept the Lags and the Tursiops apart always—we
weren't sure just how they would interact with each other, and we
would have no simple, satisfactory way of separating them once
they were together. They could potentially have been trained to
gate independently, but we had neither the time nor the need for
that now. We just kept the net gates between them, so they could
easily see and hear one another.

At first Echo and Misha shied away from the Lags, swimming
to the other side of their tank. Then curiosity got the better of
them, Echo particularly. The Lags were curious from the outset,
especially Sira—previously known as Ten. (The Lags had finally
been given names instead of just numbers. Since their aquarium
setting would depict the Pacific Northwest, the Shedd authorities
chose names taken from Tlingit, the language of certain seafaring
tribes native to coastal Alaska and Canada. Sira meant "white-
sided." Eleven, the male, was given the name Bulea, which translates along the lines of "he who swims fast." That gave us
dolphinettes something of a laugh, since he generally seemed the
slowest and least graceful of the lot. Twelve became Tique, meaning
"water born," while Nine became Kri, which still means "nine.")

Sira could hardly contain herself and spent much of her time
staring through the net at the Boys. Even at night, when the rest of
the Lags were swimming in their sleep formation, she would break
away regularly to check out those strange guys in the tank next
door. Echo returned the interest. Bulea didn't care for the competition, apparently, and he let the Boys have it with a barrage of jaw

pops—the same aggressive sounds Echo and Misha had directed at the pool vacuum early in their stay. Sira jaw-popped some, too. Echo, and sometimes Misha, fired back, though less frequently. Kri and Tique mostly stayed out of the fray.

Ted Cranford, another one of Ken's graduate students (nicknamed Grizz for his bearlike build and heavy beard), decided to take advantage of the opportunity to look more closely at this jaw-popping behavior. He'd been studying sound-production mechanisms in dolphins using modern medical imaging techniques, such as CAT scans and magnetic resonance imaging, and had identified structures that might well be the source of their echolocation clicks. (As mentioned earlier, dolphins don't produce their sounds through the larynx, as other mammals do; just where and how they *do* make their various sounds hasn't yet been firmly established.) Grizz believed that jaw pops constitute a fourth type of dolphin vocalization (along with whistles, clicks, and burst-pulse sounds), and that the sounds are not a percussive sound generated by merely clapping the jaws together. The pops might even be a version of the "big bang" sounds Ken Norris had hypothesized as a prey-stunning technique.

Now we had these dolphins popping away at each other regularly. Grizz worked with Lisa to try to "capture" the jaw-popping behavior—to get it on cue—in both Bulea and Sira. First Grizz played a tone under water, via a hand-held electronic keyboard hooked up to an underwater loudspeaker, each time either Bulea or Sira jaw-popped at Echo and Misha; Lisa immediately tossed them a fish. Thus the jaw pop, the tone, and the fish would all be associated. The idea was that once the two Lags connected tones with jaw pops, Grizz could simply play the tone and the dolphin would pop in response (and be rewarded with a fish); then Grizz could elicit the behavior whenever he wanted, rather than waiting for the Lags to jaw-pop of their own accord.

All this was happening during the last couple of days before transport. I feared that having the Lags popping at them all the time—and seeing us rewarding them for it—might be distressing for the Boys. The last thing I wanted was to upset them so close to transport time; the transport itself would be stressful enough. Besides, they probably already had a sense that something was up, what with all of us running around making preparations. I now found myself as trainer in potential conflict with a researcher. The dolphins' health was the first consideration, research be damned if need be. I told Grizz I might have to shut down his operation if the Boys started reacting adversely.

Echo and Misha didn't seem particularly perturbed. Lisa, whose sense of things delphinic I trust implicitly, thought they seemed fine with it, too. So in a matter of just three days—all the time remaining before the Boys left, and with them the natural stimulus for jaw pops—she and Grizz did indeed get Sira and Bulea to jaw-pop on cue, an impressive bit of training.

Grizz was then able to continue studying the jaw pops, recording and videotaping Sira, and particularly Bulea, in action, well after Echo and Misha had gone. He gathered some convincing evidence that jaw pops are indeed internally produced, not just jaw claps. On some occasions Bulea would pop without opening and closing his mouth. Other times, when he had been working at it for some time and seemed to be wearing down, you could see him try to jaw-pop, but he didn't have the oomph behind it—literally too pooped to pop. His jaw would open and close, but all you'd hear was a dull thud of the jaw clapping shut, not the sharp, clear pop of the real thing.

On September 10, the day before the transport, we held a press conference to let the local media know about Echo and Misha's return home and to explain the whole "sabbatical" story to them. Unfortunately, Ken was out of town at a family function, so Howard and I had to face the press. All the local newspapers, and radio and TV stations, including some from San Francisco, showed up. I'm terrified enough speaking to a group of people even without an array of microphones and cameras in my face. (No, I hadn't taken the recommended public speaking class.) Howard had to field the first several questions, while I sat there like a deer in headlights. Finally I mustered enough courage to answer a few; I couldn't really avoid it, actually, since some questions were specifically about the echolocation research. I can't remember a word I said. I was much happier once we all adjourned to the tanks so the reporters could see and photograph Echo and Misha firsthand, and I could at least talk to them one-on-one, instead of feeling as though I were in front of a firing squad.

The Boys were surprisingly well behaved for the press, considering they hadn't eaten all day. As with the transport from Florida, we didn't feed Echo and Misha at all the day before transport to ease the trip. Not only that, but they'd undergone their official pretransport physical exam that morning.

Dr. Forrest Townsend, a Florida veterinarian, was to accompany Howard and the dolphins on the transport. For political reasons it seemed advisable to have a Florida vet involved. Jay Sweeney recommended Forrest, who had prior dolphin experience and had worked with Randy previously on the Sarasota project. Forrest conducted the physical, found both Echo and Misha in good health, and certified them ready for transport.

Following the press conference, we gathered all the equipment and records that needed to accompany Echo and Misha to Florida. We also packed several of the Boys' favorite pool toys, including Buddy, the plastic boat buoy Echo so enjoyed. We thought having his Buddy with him might ease the transition to the pen at Mote Marine Lab.

Once we'd finished the last-minute preparations, all of us headed home for some dinner and a brief bit of rest. The dolphins' flight was at seven A.M. the next morning out of San Francisco, and they needed to be there about two hours early. We'd have to send them on their way from Long Lab by around three A.M.

We started gathering back at the lab about eleven P.M. I'm sure the Boys really knew something strange was about to happen when we all showed up in the middle of the night, turned on the lights above the tank, and started draining out the water. They expressed their opinions on the matter by chuffing and tailslapping.

One of the crew had picked up a Ryder truck earlier in the day, and now we backed it into the dolphin compound near the tanks. We took Misha out first and placed him in the transporter. Then we hoisted the transporter by forklift onto the back of the truck. True to form, Misha remained calm. Two people stayed with him while the rest of us went back for Echo. True to form, he protested. He struggled as he was lifted from the tank, as he was put into the transporter, and as the transporter was hoisted into the truck. He kept it up even in the truck.

Ken showed up around two-thirty A.M. to wish the dolphins bon voyage and thank the crew for a job well done. I had been feeling a bit miffed at him for not being available for the press conference (*he's* very good at these things), but I was in fact very happy to see him. His presence, his "blessings," seemed to make the farewell a bit easier for all of us.

Chris and a couple of others rode in the back of the truck with the Boys, Howard, and Forrest. Several others drove to the airport separately, to say their final farewells from there. I had to say goodbye as the truck pulled away from LML. I didn't have time to accompany them to San Francisco; I had to rush home, shower, and get ready for my own long, long flight. Fortunately, this wasn't really goodbye for me—I'd be going to Florida for the release and would see Echo and Misha then. I cried anyway. It was my last day with them as trainer, as compeer—the end of our relationship as it had been. When next I saw them, they would be well on their way to becoming wild dolphins again.

I've never been very good at sleeping on airplanes, and though I'd had no sleep that night (and not much the last few days before transport), I remained awake through most of the flight to Rome, wondering about the Boys and how their trip was proceeding. We were somewhere over the Atlantic when CNN news came on the video screen with a story from a San Francisco station. I looked up and saw Echo's and Misha's faces peering up at the camera from the LML pool, followed by a shot of me at tankside saying, "I'm going to miss them like crazy."

10

Free at Last

I called Chris from the airport as soon as we landed in Rome. It was three A.M. Santa Cruz time, but she'd said I could call anytime, and I needed to know how Echo and Misha had fared. Howard was to call Chris once the Boys made it to Florida and were safely in the pen at Mote Lab; then Chris would pass the word on to everyone in California who needed to know. Chris was groggy, but she answered the phone. She told me the truck ride to San Francisco went fine. Echo had calmed down once they had started moving, and the crew got the two dolphins loaded up and on the plane without mishap. She hadn't yet heard from Howard, though. That worried me, since we'd figured they should arrive at Mote Lab around midnight Florida time (nine P.M. West Coast time). There was nothing I could do from Rome, however, except fret—and call back later.

Joel and I found our hotel room, and I finally got some sleep. We woke up in late afternoon and I tried Chris again. She had the news I wanted to hear: Echo and Misha were fine. Howard had simply been too exhausted to call—or to remember to call—when they got the Boys into the water and settled in for what was left of the night.

Accommodations on Echo and Misha's return trip to Florida were similar to those on their original trip out to California. Again, the two people, the two dolphins in their transporters, and all their gear were confined to a ten-by-ten pallet. Howard had arranged to have the temperature dropped to fifty degrees in the cargo hold, and to have the pilots make takeoffs and landings as gradual as possible to minimize the amount of water spilled. He and Forrest tried to keep the dolphins' water at about sixty-five degrees by adding fresh water and ice as necessary. They had changed the transporter water upon arrival in San Francisco and changed it again during a layover in Memphis.

The wet-style transporters, allowing the dolphins to float in water, should have made this trip more comfortable for the Boys than their first flight. Probably the biggest difference for Echo and Misha this time around, though, was that they had been through it all once before. By now they were undoubtedly somewhat inured to the oddities of human behavior—they knew strange things could happen when people were around. I suspect the experience didn't seem quite so alien and overwhelming this time.

Howard and Forrest monitored the dolphins' respiration and heart rates every half hour. Misha remained calm throughout the trip. Echo, however, started to thrash again on the plane. Forrest injected him with Valium to quiet him. (Joel, an unhappy flier, wouldn't have minded some of the same on our flight.)

The plane landed in Fort Lauderdale at seven-fifteen P.M. Randy, Michelle, and several people from Mote Lab met the dolphins at the airport. The crew changed the transporter water one more time, loaded the dolphins into a rental truck, and headed for Sarasota.

The Boys arrived at Mote Lab around twelve-thirty in the morning. The crew decided to put Misha in the water first. They weren't sure how the dolphins would react once they were in the pen, whether they'd rush the net or try to escape. Misha was likely to be the calmer of the two, they figured, and so might have a quieting effect on Echo. At least Echo would probably stick by Misha, even if he did act up.

Howard stayed in the truck with Echo while the others carried Misha down to the dock. Before they put him in the pen, Forrest took a blood sample from Misha and checked him over for any marks. He came through the ordeal without a scratch.

Crew members on the dock lowered Misha's stretcher to a team of six in the water. Ten or twelve more people arrayed themselves around the outside of the pen. Randy and several others held Misha and walked him around the pen, showing him the boundaries and occasionally nudging his head gently into the net to remind him what netting was all about. Misha remained unruffled, not struggling at all.

Next it was Echo's turn. Once they got him to the dock, Forrest took blood from him, too, and checked out his body condition. He had an abrasion on his upper melon and a small bruise on his right pectoral fin but otherwise looked fine.

The crew lowered Echo into the water to a receiving team, who then walked him around the pen. He still had some fire left in him, even after all he'd been through the past day and a half. He fought against the restraint, diving under water repeatedly. Howard, who was holding Echo's head, got dunked with each dive. He had to wonder each time he went under, "Is he going to let me come up this time?"

Echo was by no means putting full effort into getting away,

Howard figured. He wasn't seriously trying to throttle anyone, either—you'd know it if he were. The group finally managed to swim him all the way around the pen. They turned him to face Misha and let him go. Echo surfaced right next to Misha. The two of them started swimming side by side, pectoral fin to pectoral fin, touching each other frequently and breathing synchronously.

People stayed in the water with them for half an hour or so to make sure the Boys were doing well. The two dolphins seemed stiff and moved slowly but were alert and active. The crew climbed out of the water and watched from the dock.

Michelle, always ready to feed, brought out some capelin from Santa Cruz. The Boys hadn't eaten since late Monday afternoon, and here it was early Wednesday morning. They appeared ravenous, responding immediately when the fish hit the water and consuming about a pound and a half apiece.

It was nearly dawn. Howard, Michelle, and some of the others simply slept on the dock by the pen for what little was left of the night. Echo and Misha seemed to settle in as well. They looked up at the people on the dock occasionally but weren't as responsive as Michelle expected. Instead they spent their time either on the bottom or swimming slowly near the bottom, coming up for breaths two or three times a minute—a normal, healthy rate.

First thing in the morning Michelle offered the Boys some striped mullet, dead but fresh. The aversion to mullet they'd evidenced back at Long Lab apparently had dissipated—whether because they were back in their native waters, because they were still hungry from their transport fast, or maybe simply because these mullet were fresh and the ones we'd offered them back in Santa Cruz had been frozen. Whatever the difference, neither Echo nor Misha turned up his rostrum at this meal, each of them downing some four and a half pounds of the local fish.

A fresh, fully awake crew arrived to take over the dolphin-watching duties. Howard finally was able to make his phone call to Chris and then toddle off to bed for some real sleep.

Echo and Misha's home for the next month was a rectangular sea pen, thirty-two feet by thirty-seven feet, ranging from about five to eight feet deep depending on the tide. The water temperature was in the high eighties, some fifteen or twenty degrees warmer than the tank back in Santa Cruz. Brookfield Zoo had granted Michelle a paid leave of absence so she could be with "her Boys" full-time during their final days in captivity. She was assisted by Kim Bassos and two other LML dolphinettes who had made the trip out from California, as well as by Mote Lab staff and several other volunteers.

I couldn't afford to join them, alas. I was working as a teaching assistant and couldn't take that much time off; the best I'd be able to do was fly out for the release itself. I was most unhappy not to be able to participate in this crucial phase of the venture, not to witness and assist in Echo and Misha's transition back into wild dolphins. How would they react to all the sights and sounds and smells of their home waters? Would they be reluctant to change back to their old ways, or would they eagerly cast off all things human? I had to rely on Michelle and Howard to keep me posted via telephone.

The crew kept up a twenty-four-hour watch on the two dolphins throughout their stay at Mote. At first the Boys swam mostly in small clockwise circles in the deeper half of the pen. Within a week they were covering the full perimeter of the enclosure and swimming more independently of one another, not always breathing in synchrony. By the end of their first week, both animals were noticeably "tan," their skin darkening in the Florida sunshine.

We wanted to keep Echo and Misha in the pen at Mote for a month to make sure they handled the transport well and especially to make sure there were no signs of a recurrence of Misha's illness. If something did come up, it would be far easier to deal with in the pen than if he were out swimming around Tampa Bay. Forrest prescribed a prophylactic regimen of antibiotics for both Echo and

Misha for the first week to ten days following their arrival, a common practice after a big move. It's a precautionary measure, intended to give their systems a boost to help them contend with the changes in water and in food, as well as with the strains of transport. The Boys also had to endure weekly physical exams, which involved taking blood, fecal, and stomach-fluid samples, as well as checking their general bodily condition.

The second principal purpose of the "halfway house" period at Mote was to ensure that Echo and Misha could, indeed, pursue, catch, and eat their own food. The process of getting them to eat live fish was essentially the reverse of the procedure we'd used back in Ruskin to get them to eat dead fish in the first place: first they got freshly dead fish, then incapacitated live ones, and finally fully functional live ones.

Several Mote staff members and volunteers provided the live fish. They brought in whatever they could catch, presenting Echo and Misha with a wide variety of local fish, including striped mullet, silver mullet, pinfish, redfish, spotted sea trout, grouper, grunt, sardines, shad, jack, ladyfish, and bluefish. Silver mullet was the clear favorite, however, eliciting enthusiastic reactions from both Echo and Misha once they caught on to the live-fish notion.

Our crew kept the fish in a small pen by the dock and fed them and debilitated them, as needed, to feed to the Boys. The worst part, as it had been two years previously, was maiming the poor fish—either bonking them on the head to stun them or clipping their tails.

The first time Michelle threw live fish into the pen, the Boys just watched them go by. Then they looked up at her as if to say, "Well, that's nice, but enough of the entertainment—bring on the meal."

She tried, especially at first, to keep mealtimes as familiar as possible for them. She brought the fish out in buckets, blew the whistle three times to begin the session, and asked the Boys to do various behaviors to get their fish. She tried to keep everything the same as back at LML, except that the fish were alive.

For the first week or so Echo and Misha caught and ate a few of the live fish, but mostly they seemed to regard the fish more as a source of amusement than of sustenance. They might try a bit of a chase, but nothing too serious. Michelle tried to make it into a game for them. She played off the competition between the two of them, putting a live fish between them, hoping that they'd both go after it.

That took a while. At first the people feeding had to hold the fish right in front of the dolphins to get them to eat it. The Boys managed that readily enough. Then instead of simply handing over the fish, the trainer would push it a little so the dolphins would have to lunge after it.

Within about two weeks the Boys were fairly reliably chasing down and eating fully functional fish. They even started to anticipate the toss, heading for the fish before it hit the water. Then Michelle added occasional "fish from heaven" sessions, in which she'd try to sneak up on the Boys (not an easy matter) and, without blowing the whistle to forewarn them, fling some fish into the pen. That took them a little getting used to, apparently, but before long Echo and Misha were nabbing these unannounced meals as well.

Michelle worried a little whether in the wild the Boys would go ahead and chase fish down from the beginning or still expect some person to provide them. The latter seemed unlikely, however. The crew frequently saw Echo and Misha "shushing" and "whooshing" through the water, apparently chasing the fish that got away. The night crew even reported seeing them—lit up by the bioluminescence of the plankton in the water—chasing fish in the dark. They couldn't tell whether the Boys were actually eating what they chased, but at least they had the right idea. Catching fully functional fish may be somewhat more difficult in the pen than out in the wild: though the dolphins might be better able to corner the fish in the pen, they couldn't chase them down in the same way they would when free.

Michelle made sure the Boys received a base amount of ten to fifteen pounds of fresh dead mullet apiece each day. No matter what, we wanted them to stay in good health and, if anything, maybe carry a couple of spare pounds when they were released— just a little extra to fall back on in case they had trouble. She did arrange several test days, however, when the Boys didn't get any dead fish at all—they would have to fend for themselves, tracking down the live fish in their pen.

The plan had also called for reducing the Boys' interaction with humans over the course of their stay at Mote. As it turned out, no planning was necessary—Echo and Misha made that decision for themselves.

Initially, Michelle continued to do some regular training sessions. She at least wanted a morning session in which she could look at their eyes and in their mouths and check out their general body condition.

The Boys cooperated for the first week or so. Then they became progressively less willing to come to station. If they did come over, often they wouldn't stay long—they'd just take their fish and swim off. Or they wouldn't do the behaviors asked of them. Echo even snapped at Howard's fingers during one session.

Early on Echo and Misha would spend some time playing with people. Michelle didn't want to frustrate them by cutting off their interactions with humans too abruptly, so occasionally she and Howard or another crew member would sit in a small inflatable boat and let the Boys come by to be rubbed. They seemed to like it at first. Then they started to get snippy about it, sometimes even slapping their tails on the water and making it perfectly clear, "I don't need these rubs." Michelle still needed the rubs, though.

Once after a feeding session, when the fish were gone, Michelle gave Echo the "retrieve" sign, thinking he might bring back a ring that was floating in the water. He left and returned with a fish in his mouth. He shook it and ate it in front of her—as if to say, "I don't need your fish, either."

The Boys apparently felt less need of their human-made toys as well. Both occasionally pushed Buddy around, but Echo didn't seem to have the same sexual fixation on it he'd had back at LML. Every once in a while one of them would play with a Frisbee. Eventually, they seemed to forget the toys altogether. Or they came up with their own, more natural toys. On several occasions Echo dragged up some sea grass from the bottom and swam around with it draped over his dorsal fin.

People got into the water with Echo and Misha periodically to clean the pen. Once Michelle just drifted around awhile, hoping for one last interaction of the sort we used to have with them, full of rubs and dorsal tows. The Boys ignored her. They were growing increasingly independent of people, just as we'd hoped they would—and they were breaking Michelle's heart in the process.

Echo and Misha spent more and more of their time at the far end of the pen, away from the people on the dock, looking out into Sarasota Bay. Often they could see manatees or other dolphins off in the distance. The Boys would orient toward them but generally remained silent. Sometimes, though, they would start chuffing or tailslapping after the other animals had left.

Their first close encounter with the wild dolphins of the area came after the Boys had been at Mote about a week and a half. Well before the crew noticed anything, Echo and Misha oriented themselves, facing out, in a part of the pen they didn't normally frequent. Then three dolphins, all Sarasota residents known to Randy, appeared not far away. Echo and Misha echolocated on them. Michelle heard faint whistles, probably coming from the trio. The wild dolphins didn't stop or come over to look at the Boys face-to-face, but they did pass within ten yards of the pen. The Boys followed the wild dolphins around the perimeter of the pen as they came in and left.

Two members of the trio, one known as Duckbill and her mom, came back at least three more times while Echo and Misha were at Mote. (Duckbill got her name not because of any distinguishing physical characteristics but because Bill, who was recording data at the time, was standing in Randy's way as he was trying to photograph the animal. Randy kept saying, "Duck, Bill." Bill recorded the phrase as the dolphin's name.) Whether the visits were intentional—perhaps the two sets of dolphins knew each other—or purely incidental, only the dolphins could tell.

Three more dolphin visitors, what appeared to be a mother and calf plus one single adult, showed up about a week later at three in the morning. The night was too dark for the watch crew to identify the individuals. This group came even closer than the last, within three or four yards of the pen. The encounter elicited lots of vocalizations including echolocation, burst pulses, and jaw pops. The Boys tailslapped and breached repeatedly for several hours after the wild dolphins left.

During most such visits, including one from Melba, the oldest known dolphin in the Sarasota community, Echo and Misha remained quiet, simply watching the wild dolphins from the far end of their pen. Occasionally the Boys would swim over to the dock, briefly look up at the people sitting there, and then return to watching the wild dolphins. Whether the message was one of frustration, supplication, or just a "Hey guys, check this out" wasn't clear to the humans. Once Michelle saw Echo pushing gently but insistently against the netting in the far end of the pen, apparently testing its strength. That message was clear enough.

By the time the Boys had been in the pen about two and a half weeks, they seemed clearly frustrated. They engaged in more and more tailslapping, breaching, and chuffing—noisy and presumably angry or agitated behaviors. Mote's resident caretaker, who slept on his boat at the dock, complained regularly to Michelle that "those dolphins of yours kept me up all night long."

Echo was also getting harder and harder to handle during physical exams. Misha continued to be gentlemanly about it, but Echo got a little worse each time. During the exam on September 27, he altogether refused to accept the stomach tube. He made open-mouth threats to several people and actually raked one Mote staff member slightly on the hand.

Michelle was concerned that their continued confinement was taking a toll on the Boys at this point. She and Randy talked to Jay Sweeney about the possibility of moving the release date ahead somewhat. Echo and Misha had already passed all our behavioral tests: they were catching their own food, eating a variety of fish, and showing interest in things other than people. They had passed all their physical exams as well, with blood values remaining at normal levels even after the antibiotics ceased. And the Boys were giving every indication they didn't want to be in there anymore. The reacclimation stage had been a rousing success; Echo and Misha were ready for the next step.

Jay agreed to moving the release date up several days—a total of three and a half weeks at Mote rather than the month or more we'd originally planned. Randy set the release date for Saturday, October 6.

Virtually the whole crew who'd taken part in the capture of Echo and Misha gathered again for their release—the original Santa Cruz contingent (most of whom no longer lived in Santa Cruz), Randy's Sarasota team, Jay Sweeney, the Earthwatchers, and other volunteers who'd helped out two years previously. Jill was the only major player who wasn't there. There were several new faces—Kim Bassos, of course, plus several people from Mote Lab who helped with Echo and Misha's readjustment period and who would participate in the follow-up study as well.

I had to make it a quick trip—just a long weekend—so I wouldn't miss too many classes. Casey and I caught an early-morning flight Friday. We took a cab straight to Mote Lab to see the Boys. I couldn't believe how much I'd missed them the previous month. In fact, I'd been having a hard time going to LML at all; seeing the tanks but no Echo and Misha was just too much for me. As Casey and I reached the dock, we saw the Boys bobbing in the water at the far end of the pen, staring out into the bay. I called their names, but they didn't respond. They swam by the dock a little later and looked up at me with only casual interest; they didn't seem to have missed me as much as I'd missed them. I had to keep reminding myself that this was what we wanted.

Later Casey and I met up with the rest of the gang at the Wimpy House, which served as the base camp for Randy's Sarasota project. (Wimpy is the name of the family that owns the house; the moniker isn't intended to be descriptive of the house itself nor of its inhabitants.) An Earthwatch team was there, too, so accommodations were limited, with twentysome people crowded into the place. Casey and I, late arrivals, each claimed a spot on the living-room floor, a blanket and a pillow as our bed for the night. It was a grown-up slumber party of sorts, with all of us in a high state of excitement and anxiety. I don't think anyone slept much.

I felt a little like a parent preparing to send kids off to college. Our Boys were heading out into the world, and I worried about how they would fare on their own. I very much wanted this for them but didn't want them to leave. I was happy for them, sad for me. I knew I wouldn't see much of them after this—and I couldn't even tell Echo and Misha to write or call collect. There was even some chance I'd never see them again.

We all gathered at Mote Lab at seven the next morning to get the Boys ready for their next big venture. The tide was high, so only the tallest of us got into the water to catch the dolphins; others waited on the dock to receive them. As usual, Misha went first. We did a brief, final physical checkup, weighing him and conducting an ultrasound exam. He looked good. We carried him to the RV *Mako,* Mote's forty-foot research vessel, which would convey the Boys on the final leg of their journey home, and placed him in the transporter. I stayed on the boat with Misha while the others went back to get Echo. Misha, as usual, remained calm, his eyes wide open, watching everything. Echo, as was his wont, kicked and whistled as they carried him aboard.

Randy, Michelle, Howard, Forrest, Jay, and a couple of Mote staff people went with the Boys in the *Mako* for what would be a long, slow trip out to Tampa Bay. On the ride out Randy and Jay fitted Echo with a small radio transmitter on his dorsal fin. It attached with magnesium bolts designed to dissolve in salt water, so the device would fall off within about two weeks. Echo fidgeted as they put in the bolt holes but settled down for the rest of the ride.

Our original plan had been to radio-tag both animals, but then when Misha got sick, we had to rethink the decision. We didn't want to do anything that might jeopardize his health in any way, and while the transmitter is relatively small—about the size of a Polish sausage—it might be somewhat cumbersome or bothersome to him. On the other hand, we didn't want to lose track of Echo and Misha, especially during that critical first week or two. In fact, we needed to be prepared to recapture them if it looked as though they weren't making it. We compromised: we'd put a transmitter on Echo but not one on Misha. That was something of a gamble— we were betting that the two of them would stick together so we could track them both with one transmitter—but it seemed the best solution. We equipped one of Larry Fulford's boats, the *Saida Beth*, with radio antennae for tracking.

As the *Mako* and its crew headed toward Tampa Bay, the rest of us drove back to the Wimpy House. We'd be traveling in smaller, faster boats and were leaving from a part of Sarasota Bay much closer to the release site, so we didn't need to leave until later. Those of us who'd be on the all-night tracking team packed up the food and gear we'd need for the trip. Casey and Kim Urian had both brought champagne for a postrelease celebration. We were planning on staying out, tracking nonstop (again, on a boat with no head), through Sunday night. The flotilla of about seven or eight boats, including a press boat for the media covering the event, rendezvoused shortly before noon just outside Bishop's Harbor, where we'd caught Misha two years previously.

We lowered Misha into the water first. Five of us held on to him while others brought Echo. I stood by his head, stroking his melon one last time. Once both dolphins were in the water, we oriented them offshore. It was twelve-fifteen. Randy counted down, "three, two, one . . . go!" Amid great cheering and clapping, we released the two dolphins and sent them on their way.

We couldn't crack open the champagne quite yet, however.

Echo and Misha always had a way of surprising us, and this was no exception. I guess we'd all imagined them leaping gaily off into the horizon—free at last. Instead, they circled us twice, then headed straight for the shallows, plopped themselves down on a sandbar, and lay there, dorsal fins sticking up out of the water. We'd barely shed our first tears before the dolphins up and beached themselves.

We hurried over and pushed them into deeper water. Echo and Misha just swam a short distance and landed back on the sand.

Had they grown too accustomed to having walls and a tank bottom? Were they waiting for us to take them home and feed them lunch? Or maybe they were just arguing about whose family to visit first. Probably they were trying to figure out what in hell was going on. After all, we knew they were being released; Echo and Misha did not.

We decided to stand back awhile and let them catch their breath, gather their wits, collect their thoughts, or do whatever disconcerted dolphins need to do. This time, at least, they seemed to be orienting offshore, moving their heads back and forth as though they were scanning whatever might be out there and trying to get their bearings.

After about half an hour Randy and Michelle waded out to them while the rest of us returned to the boats. The Boys rubbed against their legs. Echo rolled on his side, lifting the radio transmitter out of the water, and looked up at Michelle as if to say, "Take this thing off me"—just as she'd always done with the eyecups after an echolocation run. When she didn't do so, he tried the same maneuver on Randy. Again to no avail. Randy and Michelle walked along slowly, splashing their hands lightly on the water as they went. Echo and Misha followed. They looked like Ma and Pa, taking their dolphins for a stroll. Gradually, they led the Boys off the sandbar. We held our breath.

This time they were ready to go; this time we got our chance to cry. Echo and Misha were home.

11

Ordinary Dolphins

*E*cho and Misha headed northeast, toward Port Manatee. We followed in boats. Echo disappeared under water awhile, and signals from his radio transmitter began to falter. Maybe he was trying to rub the thing off against some rocks or an oyster bed on the bottom—a trick he'd learned with the eyecups back at Long Lab. He resurfaced, transmitter still in place, and he and Misha swam on. They soon encountered a small group of dolphins traveling in the opposite direction. There was a brief bout of splashing, and then the group continued on its way, the Boys on theirs.

Shortly after two P.M. the radio transmitter gave out altogether. So much for modern technology. Whether the failure was Echo's doing or simply some equipment malfunction, we had no way of knowing. Either way, we would now have to rely entirely on visual tracking. The transmitter still served a more old-fashioned purpose, at least—its bright orange color made Echo much easier to spot and to identify at a distance.

The day was a windy one, and before long we lost sight of the Boys in the choppy water. The boats split up and continued searching throughout the afternoon. Our all-nighter was off, since we no longer had any means of tracking the dolphins in the dark. The boat I was on didn't catch sight of them again. One of the other boats saw them about five o'clock entering Bishop's Harbor, essentially back at their starting point. The Boys swam in along the mangroves, hugging the shoreline, where the water was actually at its deepest. The boat couldn't maneuver in that close, however, and farther out the water was too shallow, so our crew was unable to follow. The first boat was joined by a second of our small fleet, and the two boats waited just outside the harbor mouth, in case Echo and Misha should venture back out. The crew watched for nearly two hours. No sign of the Boys. The two boats headed for home just before sunset. Bishop's Harbor, a quiet, dead-end bay with a plentiful supply of fish, apparently offered a safe haven to a pair of bewildered dolphins on their first night back in the wild.

We headed out again early the next morning, Sunday, but none of the boats saw either Echo or Misha that day. Casey and I had to leave the following morning, so we'd seen our last of Echo and Misha, at least for the time being. Other members of the team tried to go out that day, but the winds were too high. By Tuesday the high winds had turned into Tropical Storm Marco. Small boat travel was now impossible and remained so for several more days, preventing the follow-up team from continuing the search.

On that same Monday a woman who'd read the news reports about Echo and Misha called in from Ruskin to report seeing a couple of dolphins—one of whom had a bright orange gadget on its dorsal fin—apparently feeding on mullet near shore. Later that week a local TV station offered helicopter search time in exchange for an interview. Randy agreed. He and Howard went up in the helicopter. Howard very nearly lost his lunch, but they saw no sign of the Boys. The helicopter newsman did report seeing Echo and Misha several days later, however, Echo's gaudy transmitter again providing the identification.

Finally on Friday, October 12, six days after the release, the follow-up boat could venture out again. The crew found Echo and Misha within a few miles of the release site. They were on the outskirts of a larger group of dolphins but didn't appear to be a part of the group. Echo still wore the radio transmitter, though it had lost both antennae. The Boys approached the boat and rode the bow for about fifteen minutes, giving the crew enough time to check out their body condition. No indication of dehydration or weight loss. No undue number of rake marks from other dolphins. They looked great—like normal, healthy wild dolphins.

Despite the early failure of Echo's radio transmitter, the follow-up study was off to a good start. This aspect of our dolphin "sabbatical" program was completely unprecedented. Others had freed captive dolphins before, but as far as what happened to the animals afterward, they had, at best, only a few casual reports from fishermen or boaters. No one had ever carried out the sort of long-term, systematic follow-up we were undertaking.

Kim Bassos was largely responsible for conducting the follow-up study and for collecting the data, which would serve as the basis for her master's thesis. Howard and then Holly filled in for her during the winter of 1991, when Kim had to return to Santa Cruz for course work. Various volunteers and staff people from Mote Lab also took part. Randy and Michelle joined Kim on occasion, as did several others from the LML dolphin crew.

Kim's research would be twofold, both aspects aimed at assessing Echo and Misha's reacclimation. First, she would compare the Boys' behavior in captivity to their behavior in the wild. Kim had originally assisted Jill in the behavioral observations back at Long Lab and then took over in April 1990, after Jill left the project. Specifically, she looked at the dolphins' surfacing positions (did they surface for breaths synchronously or not, and if so, what position were they in relative to one another?), activity budgets (how much time did they devote to each of various activities?), and dive durations (how long did they stay under water at a time, and how did that vary with activity?). She observed the Boys not only at LML but also in the pen at Mote Lab. Now she would see what they did back home. For the first time we could compare the behavior of the same two dolphins in these two very different environments—in our world and in theirs.

Second, she would compare Echo and Misha's behavior to that of other dolphins in the area. Here she would look at their ranging patterns (where and how far they traveled in the normal course of their day-to-day lives) and their association patterns (who they hung out with and how often). Would Echo and Misha return to a normal dolphin social life, or had their stint in captivity made misfits of them?

The research plan called for an intensive year-long follow-up study, after which the study would continue opportunistically, with a crew going out to look for the Boys whenever people, time, funds, and boats were available. Kim and crew would search for the dolphins three to four days a week, weather permitting, for this initial year. For the first month she used a boat donated by Cannon's Marina in Sarasota. After that Mote Lab provided the *Minimako,* a nineteen-foot outboard with a 130-horsepower engine.

Kim was relatively new to fieldwork. She had helped the preceding summer on Randy's dolphin survey of Charlotte Harbor, just south of Sarasota Bay, but she was still basically a novice. She had to learn not only the survey and dolphin-identification techniques, but also how to drive the boat, how to navigate Tampa Bay with its tricky shallows, how to read and contend with its waves, tides, weather, and so forth. Pete Hull, operations manager at Mote

Lab (and the primary person who had provided Echo and Misha with live fish in the pen), was one of several people who helped teach Kim the workings of the boat and the bay. He romanced her while he was at it, courting her atop the sighting tower—its platform built for two—of his boat, the *Hoog*. They married a year and a half later, with Randy officiating, as he had for Kim Urian and Andy Read. (At about the same time, Howard also became involved with another Mote employee who worked on the follow-up study, making a total of four couples brought together, at least in part, thanks to Echo and Misha.)

On a typical day Kim and crew would set out from the Wimpy House at about eight or eight-thirty A.M. Generally the team consisted of three or four people, plus the occasional dog. The ideal crew number was four: once they found a group of dolphins, they would have one person to drive the boat, one to photograph the animals for identification purposes, one to record all the necessary information on data sheets, and one simply to watch, to keep an eye on where the dolphins were and what they were up to. Sometimes, though, Kim went out with just one or two other people; other times as many as six people were on board. Once Kim even ventured out alone.

They concentrated the search effort along the southeast coastline of Tampa Bay, about a twelve- or fifteen-mile span around the release site. Generally, they focused on the near-shore waters between the Skyway Bridge to the south and Ruskin, where Echo and Misha had been temporarily housed following their capture, to the north. Occasionally, however, Kim would do zigzag transects a little farther offshore or cut into Terra Ceia Bay and the Manatee River, south of the Skyway. During the annual fall-census surveys (1988–93), and occasionally at other times of the year, the search covered all parts of Tampa Bay.

The crew would load up all their gear—cameras, data sheets, video camera, occasionally a hydrophone, plus their lunches and plenty of sunscreen—storing it all in an ice cooler to keep everything dry. As they drove, everyone on board would search in all directions for signs of dolphins. If the animals are actively socializing, leaping, and sending up sprays of water, they may be relatively easy to spot from a distance. Otherwise, it can take keen eyesight, especially if the water is rough and the dolphins are moving fast. The crew scanned all around the boat, looking for any sort of break in the surface or disturbance in the water. They also looked for seabirds hovering or diving in a particular area. Frigate birds are an especially good indicator of dolphins below. Notorious for snatching food from other birds in midflight, frigate birds aren't above stealing fish from a careless dolphin either. If somebody spotted one, Kim would often go off course, aim toward the bird, and check out under it. More often than not she'd find dolphins there.

They almost always saw some dolphins when they went out, whether they ran across Echo and Misha or not. During the entire follow-up period Kim got skunked only once. That was a nasty, windy day, so it was hard to see much of anything. The crew occasionally would brave such a day after a bout of bad weather had kept them in for three or four days in a row and they were desperate to get out on the water again. That was Kim's least favorite part of the entire operation—pounding their way across Tampa Bay with the wind blowing at fifteen knots out of the northeast, getting soaking wet, and then not even finding any dolphins, much less Echo or Misha.

When they found a group of dolphins, they would first collect all the basic information Randy used in his Sarasota surveys. That involved stopping and getting cameras out, photographing the animals, and recording the roll numbers and exposure numbers on the data sheet. They would record the date and time they sighted the dolphins, location, latitude and longitude readings, environmental variables, such as chop, wind, glare, tidal conditions, water temperature, and depth, the dolphins' initial and general headings, their activities, and the group size, including the number of calves and the number of "yoys" (rhymes with "boys"), which stands for "young of the year."

If they saw Echo or Misha, they would fill out the basic data sheets and then Kim would start her "focal-follow" study, "sampling" their behavior and group composition once every five minutes for as long as the boat could stay with them. She also recorded dive durations, along with activity at each surfacing, whenever possible, and kept a continuous record of activity, location and depth, group size, and composition. Someone on the crew generally videotaped the encounter to assess the Boys' body condition and to record both typical and unusual behaviors.

During focal follow the boat paralleled alongside the dolphins at a distance of about ten to thirty yards. Kim tried never to drive into the midst of a group, since she wanted their behavior to be as natural as possible. She generally wouldn't start the focal follow until fifteen or twenty minutes after they approached a group, allowing the dolphins time to get used to the boat first.

Sometimes she would be able to stay with the group for only ten minutes. Her record follow was more than four hours. Usually the dolphins would let her know when they'd had enough. If they started tailslapping a lot, chuffing, or being very elusive, Kim said, "Okay, time to leave." Some days the dolphins tolerated the boat and its people with equanimity. Other days their attitude seemed more one of "Let's lose these guys."

The second official sighting following release came on October 18. Again the Boys were about one hundred yards from a group of dolphins but not actively engaged with them. Echo and Misha approached the boat and rode the bow.

Four days later Kim and crew found them in a group with

other dolphins for the first time. The Boys interacted most extensively with two others, one with a very raggedy trailing edge of its dorsal fin and the other with a white scar on the leading edge of its fin. The two were dubbed Bumpy Fin and Scartip, respectively.

Echo, Misha, and their two friends approached the boat and went bow-riding together. All four dolphins kept turning at the bow and looking up at the people—who were leaning over the bow, yelling Echo's and Misha's names and saying, "Hi guys, hi guys." Sometimes they came so close they actually bumped up against the bow, rubbing against the side of the boat, which caused the humans some concern. The boat was traveling slowly, however, and the dolphins seemed to know what they were doing.

Then a tanker passed by in the shipping channel, sending out a big wake, which got even bigger as it hit the shallows where the dolphins were. Misha, Bumpy Fin, and Scartip broke off their bow-riding and instead went surfing on the tanker wake, leaping and splashing and seeming to have a high old time. Echo stayed back, still swimming alongside the follow-up boat.

Echo's reticence gave Kim pause—why wasn't he out surfing with the others? That seemed uncharacteristic. He still wore part of the radio transmitter, however, and he might have felt too encumbered by that for any serious acrobatics.

By the next encounter, on Halloween, the last of Echo's transmitter apparatus had dropped away, leaving only small scars where the bolts had been. The Boys were together but without any other dolphins in sight. That was the last time they were seen by themselves. From then on they were always in the company of others, Bumpy Fin and Scartip among the regulars.

As of November 4 the follow-up crew had put in twenty-one days of search time, sighting Echo and Misha five times after the release day. During that first month of freedom the Boys were always together, initially on the periphery of dolphin groups but then slowly mixing, becoming part of the community. By the end of the month the crew had observed them doing all the normal dolphin activities—traveling, milling about, resting, feeding, socializing with other dolphins, bow-riding, leaping. They also appeared to be maintaining excellent body condition.

Then on November 5 the crew found Echo with two mom-and-calf pairs just south of Cockroach Bay. No sign of Misha. Kim started to worry—"What happened to Misha?" The very next day, however, she spotted Echo and Misha together again, with four other dolphins, not far from where they had found Echo the day before. Apparently the Boys were now feeling secure enough to leave each other's side from time to time. Kim continued to find the two of them together on most sightings, but it became increasingly common to run across just one of them with other companions.

Once they found the Boys amid a very active, rowdy group of dolphins. Misha kept making fart-type noises—a pastime he'd engaged in from time to time back at Long Lab as well. A couple of the other dolphins stopped what they were doing and stared at him. Perhaps the noise is considered rude in proper dolphin society, too. Or maybe it was some bad habit he'd picked up in captivity.

In early December a good friend of Kim's, also a Long Lab dolphinette, joined her for a couple of weeks. One day just the two of them went out. They didn't see Echo or Misha, but they did find their buddies Bumpy Fin and Scartip, who came over and started bow-riding. Kim and her friend decided to sing Christmas carols to them, just to get into the Christmas mood. As they hung over the bow singing "O Come All Ye Faithful" and "Jingle Bells," Bumpy Fin and Scartip rode along in the bow wave, repeatedly turning and looking up at the serenading humans. Interspecies communication comes in many forms.

Randy and Michelle joined the follow-up crew for a few days in mid-December. When they saw Echo (no Misha), Michelle, always the trainer, couldn't resist pulling out her training whistle and giving three toots—the signal we'd used to begin a training session. Echo leaped out of the water twice in a row, right next to the boat. They saw Echo (again no Misha) a few days later, too. Michelle tried the whistle. No response.

Kim left to head home for Christmas and then back to Santa Cruz to take classes winter quarter. Nobody went out to look for the Boys for a couple of weeks around Christmas, and Howard wasn't due to arrive to fill in for Kim for a couple more weeks. Misha hadn't been spotted since fairly early in December, a matter of some concern. So on January 4, Pete Hull went out searching. Just as he drove under the Skyway Bridge, the first dolphin he saw was Misha, accompanied by two buddies. Again, we had been worrying unnecessarily—the Boys seemed to be taking care of themselves just fine.

Howard arrived, staying from mid-January to mid-February, followed by Holly from mid-February to early April. Foul winter weather severely limited the number of boat-days, and sightings of Echo and/or Misha were relatively few during that time. Less familiar with Kim's focal-follow procedures, Howard and Holly were concerned primarily with just keeping track of where the Boys were and with whom.

Howard had only two sightings of the Boys during his stay, both in late January. Holly's first sighting of the Boys didn't come until March 12. Echo and Misha were together within a group of a dozen or so animals, though the two of them appeared to be in separate subgroups. Misha came over to the boat several times and looked up at Holly. He showed up four more times before she left in early April, always with other animals. She saw no more of Echo.

Holly was surprised that every time she encountered Misha in the field, he came up to the boat, rolled over, and looked up at her. Of course, she was yelling, "Misha, Misha." But he would look right at her, as if to say, "What the hell are you doing here?" The other dolphins seemed to her to be more habituated to the boat, less likely to approach it, more likely to ignore the people. Not Misha. He would even stop what he was doing at the time and leave the other animals to go look.

Kim returned in late March. Six months had passed since Echo and Misha's release. During that time various crew members had accumulated a total of seventy-six boat-days of searching. They saw Echo twenty-eight times, Misha twenty-four times; on twenty of these sightings the two of them were together. What Kim was finding already was that the Boys' behavior patterns were very similar to those of other wild dolphins. Their reintroduction appeared to be eminently successful. Echo and Misha were doing all the normal dolphin things; they were out there behaving just like ordinary dolphins.

Kim found that Echo and Misha didn't surface synchronously for breaths nearly as often in the wild as they had either at Long Lab or in the pen at Mote. This may reflect the more wide-open spaces of Tampa Bay—they had more opportunity to be farther apart from each other. They also had more dolphins to associate with in the wild, so they could well surface and breathe in time with someone else when they weren't doing so with each other.

When they did surface synchronously, however, their relative positions remained constant. Echo generally surfaced to Misha's right and slightly ahead of him at all three locations: LML, Mote, and in the wild. The more forward position is often considered one indication of dominance, so this arrangement presumably says more about the relationship between the two of them than about their captive-versus-wild status. Apparently Echo carried his domineering ways back home with him.

Kim found some dramatic changes in how the Boys spent their days. In captivity their most frequent activities had been milling (swimming in undirected fashion) and resting. In the wild they devoted a much greater share of their time to traveling (directed swimming), foraging, and socializing (any kind of interaction with other dolphins). This makes good sense. Captive dolphins have less room to travel in, and perhaps a greater luxury to rest as they choose than do their wild counterparts. In fact, Kim very rarely observed resting behavior among the wild dolphins during daytime hours. (She wasn't able to observe their nighttime behavior in Tampa Bay, so she couldn't compare that to Echo and Misha's nights back at Santa Cruz.) Wild dolphins have to travel farther to find their food and their companions; they aren't handed dead fish and so have to spend more time searching for their meals. They also have many more potential social partners to choose from than the Boys had in captivity. The changes may seem predictable. Nevertheless, Kim's findings indicate that Echo and Misha didn't maintain their captive-learned habits once they were home again, but reverted readily to behavior appropriate for wild dolphins. Even their dive durations—on the order of twenty to thirty seconds on average, both in captivity and in the wild—were on a par with those recorded for wild dolphins in the area.

To assess their social lives, Kim first determined Echo and Misha's ten most frequent companions, as identified from the photographs taken at each sighting. As it turned out, six of the ten had been previously caught and tagged as part of Randy's Sarasota project, so she already knew their sex and approximate age. (The Sarasota dolphin population and the one in southeast Tampa Bay overlap some; apparently some of the Tampa group wandered far enough south at some point to find themselves in Randy's capture-release program.) Three of the known six were males of roughly Echo and Misha's age, two (including Bumpy Fin) were females about the same age as the Boys, and one was a female several years older than the rest. Scartip, one of the unknowns, appeared to be about the same size as Echo and Misha and so probably was of comparable age.

Kim then computed what's called the coefficient of association

(COA) for each relationship among Echo, Misha, and their ten associates. This essentially denotes the percentage of time that two given dolphins were seen together, out of the total number of times either of two was sighted. A COA of 1.0 means the two dolphins were always together; 0.0 means they were never together. Echo and Misha thus far were each other's most frequent companions, with a COA of 1.0 for the first month and 0.71 over the first six months— a figure comparable to what Randy has found among male pair bonds in Sarasota Bay. Bumpy Fin and Scartip headed the rest of the list, with COA's from 0.34 to 0.40 all around, but the Boys had plenty of other friends as well. They hung out with a variety of dolphins in ever-changing groupings but maintained a regular cadre of "best buddies," just like the other dolphins. Echo and Misha seemed to fit right back into wild dolphin society. No misfits they.

Both Misha and Echo had remained within ten miles of the release site, ranging over an area that extends along seventeen miles of the southern shore of Tampa Bay. The distribution of their sightings during this time was very similar to those of their ten closest associates. All the animals, Echo and Misha included, seemed to spend most of their time between Cockroach Bay and Port Manatee, a roughly six-mile stretch that appeared to be the core area of these dolphins' home ranges. The groups' home-range areas averaged just over thirty square miles. Misha's range fell slightly above the average. Echo's range, about twenty square miles, was the smallest in the group, perhaps because he was a newcomer and didn't know the area as well.

For all intents and purposes, Echo and Misha were now behaviorally indistinguishable from their compatriots. Perhaps their only residual distinguishing feature was a somewhat greater than usual tendency to approach the follow-up boat and peer at the people in it—particularly in Misha's case. Not that other dolphins don't do the same, but they don't appear to do so quite as often. That doesn't necessarily mean Echo and Misha are inclined to approach boats or humans in general; as far as we know, it's only when the boat contains someone whose voice the Boys are likely to recognize yelling out their names. Even then, they have made no attempt to beg for food or solicit any contact beyond eye contact.

After Kim's return the weather let up, permitting more frequent boat searches. She continued to find Misha in all the usual places, with the usual cast of characters. No sign of Echo, however. Holly's March 12 sighting was the last record we had of him.

Misha now seemed to have a new best buddy. He hung out most frequently (COA of 0.60) with a dolphin called Davis, a male a year or so younger than Misha, with whom he had associated relatively little while Echo was around. Davis even seemed to have taken over Echo's old surfacing position, to the right and slightly ahead of Misha.

Months went by and still no Echo. We started to worry. Dolphins—especially young, rambunctious male dolphins—face plenty of potential hazards out there, both natural and human. His reacclimation to the wild seemed so good, so complete, however, that if something had happened to Echo, it was unlikely to be a result of his stay in captivity. That thought was of small consolation. Echo was a friend, part of the family; we simply wanted to know that he was well and happy.

In mid-September, as the anniversary of Echo and Misha's release approached, I headed for Florida to take part in the final weeks of the intensive one-year follow-up program. Echo remained among the missing.

12

The Wild Life

I timed my trip to coincide not only with the ending of the formal follow-up study but also with the annual Tampa Bay dolphin survey when the most extensive and intensive search would be under way. I badly wanted to see both Echo and Misha again, and I wanted to get some feel for their lives back in the wild.

Randy had a series of Earthwatch teams to assist in the survey project. I stayed at the Wimpy House with the Earthwatchers. Kim Urian and Andy Read were staying there as well, working on the survey and serving as "house parents" for the Earthwatchers. Kim Bassos had returned to Santa Cruz for a brief visit.

My fears for Echo only mounted after I arrived in Florida. One morning early in my stay Randy got a call from someone at Mote Lab asking for our help hauling in two dead dolphins found floating in Sarasota Bay. The marine patrol towed the carcasses behind their boat. We hoisted them up into the back of Mote's pickup truck. People at the lab would then perform necropsies on them to try to determine the cause of their deaths.

The first dolphin was a great big male, very stiff with rigor mortis—including a very stiff, and very large, erect penis. We couldn't tell the sex of the second one because much of the lower part of the body had been eaten away by sharks. The dorsal fin, the principal means of individual identification, had been bitten off as well, though I can't imagine why a shark would go after a part with so little meat on it. The shark bites and teeth marks came in several sizes, suggesting that a variety of shark were involved. The sharks, however, may have had nothing to do with the dolphin's death in the first place, but simply taken advantage of the situation after the fact.

Randy said that made nine dead dolphins in the area within the past month. Four or five more came in over the course of the next week and a half. The usual total for a whole year is only eighteen in the area covered by the Mote Lab stranding network (which cares for sick or wounded dolphins and manatees that strand themselves, as well as dealing with the dead bodies). Just what had caused the deaths—most of them, anyway—remained uncertain.

The red tide was one possible factor. Red tide is caused by tiny oceanic plankton called dinoflagellates, which are normally present in the water at some level but at times appear in numbers great enough to tinge the water red. They produce a toxin that can accumulate in the marine organisms that eat them. That's what can make shellfish poisonous to humans at certain times. The toxic effect gathers force as it moves on up the food chain, accumulating in fish, which the dolphins—top of the food chain—eat in turn.

There was a nasty red tide when I was there. You could smell it in the air. When I walked on the beach, I could hardly breathe because of the toxins. On one such walk I counted two dozen dead fish on the beach within the span of a fifteen-minute stroll.

One of the deaths was of known cause—clearly human and clearly deliberate. The dolphin had been stabbed. The weapon, whatever it was, went in below the dorsal and straight into a ventricle of the heart. Undoubtedly, death came quickly. Most likely it was done while the dolphin was bow-riding.

That death was a particularly hard one for Randy. He had known the dolphin since she'd been born—September 27, 1980, to be precise. He had seen her mother, calfless, on September 26, and then the next day with an obvious newborn. Randy named the calf Autumn because she was the first one he knew for sure had been born in the fall rather than in May to July as most of them are. She'd been found dead the day after Labor Day, 1991. Three days of the year seem to bring out the most jerks in boats, Randy said— Memorial Day, the Fourth of July, and Labor Day. He offered a thousand-dollar reward for information about Autumn's killer, but to no avail. An anonymous phone caller claimed to have been on the boat at the time of the stabbing but refused to say more because he feared for his own safety.

In this case three deaths may actually have been involved. Autumn had been pregnant, carrying a month-old fetus. And she had a fifteen-month-old calf, probably too young to survive on its own—usually they don't leave their mothers until somewhere between three and six years of age. The calf had been sighted several times since its mother's death, and it looked healthy. Apparently the youngster was managing to feed itself, though if the mother had been alive, it would probably still be suckling as well as eating fish. The calf followed with its mother's group but seemed always to be on the periphery, not really associating closely with anyone. None of the other dolphins appeared to have "adopted" the orphan, though its own grandmother was also a member of the group.

I had to wonder if there was some mad dolphin slasher out there, on top of whatever disease or environmental factors might be killing inordinate numbers of dolphins. Had Echo also fallen prey to one or the other?

If Echo had died since Holly last saw him back in March, we wouldn't necessarily know it. Not all dolphin carcasses wash ashore, or they may end up onshore in some obscure, uninhabited little inlet and never be seen. Or the body could be so badly decomposed or so shark devoured as to be unidentifiable. Had Echo survived all the trials and tribulations of life with humans only to meet an untimely death in his own backyard?

My first day out on the water, Monday, September 16, we saw several of Misha's buddies, but no Misha. No Echo either. The next day got off to an inauspicious start. I headed out in the *Makila* with Kim Urian, Andy, and one of the Earthwatchers. We were about to pull out of the channel into the bay, but when Andy tried to put the boat in gear, nothing happened. We had to call one of the other boats and have them tow us back to the dock. We took off again in the afternoon, this time in the *Minimako*. Andy figured I was the curse and something else would go wrong. Not so, though the afternoon had turned windy, raising whitecaps on the water. We were about ready to call it a day but decided to take one last swing by the Skyway Bridge first. We spotted a group of dolphins under the bridge and went to check them out. One of the group looked like Misha.

Misha, apparently, hadn't found life entirely Edenic since his return home. A couple of months earlier, whether in a tussle with another dolphin or in some sort of accident, he'd torn a flap of his dorsal fin just below the notch where he'd once been tagged. The tear considerably altered the profile of his fin—one of his primary ID cards, for our purposes. Fortunately, the follow-up crew had recognized him anyway and was able to trace the change photographically, from the initial stages of the wound through its healing.

The Earthwatch volunteer was the only one of the group who had seen Misha recently. Kim, Andy, and I hadn't seen him since the release. Kim and I had looked through some slides the night before, however, to familiarize ourselves with his new fin. We all thought this dolphin was Misha but couldn't be sure.

We followed the group under the bridge. It was even rougher and windier on the other side, but we managed to stay with the group. Several animals rode the bow and even peeked up at us. The one we thought was Misha seemed the most boat wary of the lot; he stayed under water longer and generally surfaced farther from the boat than the others. I was greatly disappointed, especially after Kim Bassos and Holly's reports of Misha's people-watching inclinations. Maybe this wasn't Misha. Or maybe he'd simply had enough of us humans.

Finally the dolphin we'd been eyeing came over and swam right by the bow. I was hanging over the front of the boat, calling Misha's name. The dolphin rolled on his side and peered up at me with one large, expressive eye. Misha. I caught sight of the freeze-brand—202. Definitely Misha.

He was in a group of at least eight animals, including several of his closest associates—Left Bend Mid Nick, Mid Scoop Low Box, Freeze-brand 28—plus an unknown smaller animal who was leaping a lot. Kim thought one of the others might be Pecan Sandy (named for the cookie, part of the standard lunchtime fare on Randy's Earthwatch projects), also one of Misha's regular companions.

I was thrilled to see Misha back home with his buddies. I was also a bit disconcerted. I wouldn't have recognized him, certainly not just from the dorsal fin, if I hadn't studied the slides first. This focus on dorsal fins felt strange to me. I was more accustomed to recognizing dolphins by their faces, the whole body shape, how they move. Michelle understood. She'd been living with this difference between the trainer's world and the field researcher's world, her world and Randy's. Each gives such a different view of the animals, of their lives. Yet the two views can strengthen and inform each other. Those of us who worked with Echo and Misha back in Santa Cruz undoubtedly see them rather differently, even when they are out in Tampa Bay, from the way people do who have never worked with them, touched them, tried to figure out what was going on inside their heads. I suspect, too, that Echo and Misha see us—and perhaps people more generally—in a different light than do dolphins who hadn't lived two years among the humans.

We went out on the water nearly every day, weather permitting. Someone always turned on the weather channel—practically a permanent fixture in the Wimpy House—first thing in the morning to see what was in store. We planned our days around these television oracles.

We saw dolphins every time we went out, though we didn't see anything more of Misha. No Echo, either. Each trip out showed me something a little different about the life to which Echo and Misha had returned, however. The dolphins there seem to keep busy and varied schedules. Sometimes they are in large groups, sometimes small; sometimes they are boisterous and playful, sometimes intently focused on finding a meal, swimming purposefully toward some destination, or just lollygagging about, taking it easy.

Once we saw a lone mother and calf out in fairly deep water, a potentially dangerous situation. A Sarasota female known as Hannah, apparently something of a loner who rarely swims with other dolphins, lost her 1989 calf to a shark. More often mothers with calves will travel together in larger groups, providing mutual protection and even "baby-sitting" for one another. Sometimes the moms swim in a U-shaped formation, forming a "playpen" within which the calves can safely romp. At least in Sarasota the females with young calves also tend to favor shallow-water "nursery" areas, where sharks are less prevalent.

Another time we saw a group of about ten or twelve animals, all mothers with calves of various sizes. One young calf had a remora, or "suckerfish," stuck to its flank and kept leaping repeatedly, presumably to rid itself of the hitchhiker. (One explanation offered as to why whales and dolphins leap is to shed parasites and the like. Leaping may also play a communicative role. But, as with bow-riding, that doesn't necessarily mean it can't be fun as well.) After a few minutes the calf apparently succeeded in dislodging the pesky fish. Shortly thereafter a nearby adult, perhaps the calf's mother, appeared with a remora on her side. Mom seemed less than pleased, launching into an extended bout of tailslapping.

Often we ran across what Michelle referred to as a "poopy group"—difficult to track and even more difficult to photograph. Generally such groups were busy feeding, with little interest in boats or people. The animals were widely dispersed, with as much as one or two hundred yards between individuals. They stayed under water for long periods and then surfaced far away from their previous surfacing, making it nearly impossible to predict where one would be at any given moment. Even more common, and more frustrating, are the disappearing dolphins—you see them once and then they're gone without a trace.

My favorite was a group of about eight animals, probably sub-adults, engaged in all manner of play and sexual behavior. After a while they split into subgroups of three and five. The group of three then started a twenty-minute round of beak-to-genital propulsion—not just a captive-dolphin propensity, clearly. One would roll over on its side at the surface and another would stick its rostrum into the first one's genital slit and off they'd go. They were extremely rowdy about it, jostling each other and seemingly vying for position—"my turn, my turn." Randy estimates that such sexual play is about equally divided between same-sex and opposite-sex encounters.

On Saturday evening, September 21, at about eleven-thirty, as I sat in bed reading, I heard a soft knock on the bedroom door. Kim opened the door hesitantly, apologized profusely for disturbing me so late, and then, unable to contain herself any longer, announced, "I've found Echo!"

Kim Urian did much of the photo ID work, both for Sarasota and Tampa Bay dolphins. She had been going through slides from survey trips earlier in the month, and there he was, in a sighting made September 3 up in Old Tampa Bay. No doubt about it. Not only did the dorsal match up perfectly, but on at least one picture you could read his freeze-brand—204. You could see even small scars from the bolts used to put the radio transmitter on him. There was one nice face shot of him, too—definitely an Echo face, with the backgammon-board forehead and the white eyeliner. What's more, he had his mouth wide open and seemed to be landing on top of some other dolphin. That's our Echo. Michelle said later she was glad to know he was a mouthy brat in the wild, too—not just with us.

One shot showed him leaping, giving a good view of most of his body. He looked in excellent shape. He was part of a group of about eight subadults, all of them being extremely rowdy. One of the others in the group was a dolphin called Small ET (to distinguish him from another dolphin they'd previously named ET), who'd been part of Echo's group when we'd captured him back in 1988.

The sighting was just north of Gandy Bridge, about twenty miles from the release site but only a couple of miles from where Echo had been caught. Apparently, he'd simply gone home. We'll never know just when or how that happened. The survey data have indicated several separate dolphin communities inhabiting different areas of Tampa Bay, including Misha's group along the southeast coast, and the group Echo returned to along the northern end of the bay. Apparently the two ranges overlap in the area around Ruskin. Perhaps Echo ran into some of his old buddies at the Bahia Beach one day and decided to head back with them. Or maybe he wandered far enough north to recognize where he was and just kept on going for home. Maybe he even knew all along where he was but needed the time to reacclimate before making the journey.

Kim and I debated briefly about calling Randy and Michelle so late and waking them up but decided this news was worth being disturbed at any hour. They agreed. Randy answered the phone, then started telling Michelle solemnly, as though it were bad news: "I don't know how to tell you this . . ." She nearly kicked him out of bed. I called Howard in Santa Cruz and asked him to pass on the word to Kim Bassos as well. I called Joel, too, and asked him to relay the message to Casey. Michelle later sent out a stack of postcards to everyone else she could think of who might want to know, using cards Mote Lab's gift shop had made at release time, with the words "Hi—I'm Echo" above a picture of his smiling face.

Kim berated herself repeatedly for not having recognized Echo at the time, when she was there taking the pictures. They had been at the edge of a storm at the time and were forced to rush the job. Still, it does show just how hard it is to identify dolphins in the field, no matter how well you may know them. Both Michelle and Kim said the only ones they can recognize readily are ones whose slides they have pored over ahead of time. They have to see the dorsal clearly, not in motion, to be able to pick it out when it just breezes by as part of a living, moving animal. Again I was struck by the differences between dealing with dolphins in the wild versus in captivity, by how different even Echo and Misha looked in the bay as opposed to the lab.

Two nights later Randy and Michelle called a "staff meeting," so Kim, Andy, and I left the Earthwatchers to their own devices and joined them at their apartment. They broke out a couple bottles of champagne, and we all celebrated Echo's reappearance. Echo was alive and well and living in Old Tampa Bay! "O frabjous day! Callooh! Callay!" we chortled in our joy. I don't think any of us had quite realized just how worried we'd been getting.

Both our Boys were home. Maybe not together, but home, each

with his own life. Apparently the pull of clan, old friends, home range, all that he had grown up with, was even stronger for Echo than whatever bond he and Misha had with each other. Perhaps they'll meet in Ruskin now and then and reminisce about old times.

About a week later, after one Earthwatch team had left and before the next one arrived, Michelle, Kim Urian, and I, together with Sue Hofmann, a Mote Lab employee who had been a regular part of the follow-up crew, decided to head up to Old Tampa Bay to see if we could find Echo, live and "in dolphin." That was about a two-and-a-half-hour boat trip from the Wimpy House, and not an easy trek—one of the reasons we'd originally chosen the southeastern coastline as the capture and release site. (We had hoped to catch both dolphins in that area; it just didn't turn out that way.) Misha's home range was not only conveniently close to the dolphins in Randy's Sarasota study, but also to his field station, the Wimpy House. Echo's home area was far less accessible.

It also seemed less inviting, or at least more highly developed. That had been another deciding factor in selecting what turned out to be Misha's turf—it's largely undeveloped, much of it still with its natural mangrove-lined shore. There isn't even much fishing along there.

We didn't see many dolphins at all, much less Echo. After searching the area fruitlessly awhile, we anchored near shore and ate our lunches, hoping maybe the dolphins would come to us. No such luck. We gave up and headed toward home around midafternoon.

The wind in the main part of Tampa Bay had picked up considerably, and storm clouds were building in the east. It quickly turned into the bumpiest, most harrowing ride I'd yet experienced in Tampa Bay. The waves hitting the front of the boat sent up a spray that thoroughly drenched us all. We were bouncing so badly that sitting down was impossible. Even the soles of my feet hurt later from the impact. What's more, we were running out of gas. Michelle started singing "The Wreck of the Edmund Fitzgerald."

We debated whether to stop off in St. Petersburg to get gas, thereby prolonging the trip, or simply hightail it home, hoping the gas would hold out. Sue, more experienced with Tampa Bay than the rest of us, recommended the former. A good thing. We probably didn't have enough gas to make it back, given that we were fighting the waves. Getting stranded in the middle of Tampa Bay amid high waves and with a storm coming wasn't a pleasant prospect. We finally pulled in to the Wimpy House, wet and weary, a little before six o'clock to find Randy and Andy anxiously awaiting us on the dock.

I left Florida having seen Misha just the once and without having seen Echo at all. At least I knew both were alive and well. That was enough. I felt a bit like the narrator of *The Little Prince*, Antoine de Saint-Exupéry's classic children's book: for him the stars would never look the same; for him the stars would always sing, since he knew his beloved Little Prince was out there among them. For me Tampa Bay would never again be the hostile place it had seemed back at capture time. Echo and Misha lived there, lending new beauty and depth to its murky waters.

The intensive year-long follow-up search ended in October of 1991, not long after I left. That didn't end our follow-up efforts, however. The annual fall census sponsored by the National Marine Fisheries Service continued through 1993, giving our crew a chance to scour Tampa Bay for Echo and Misha while surveying the rest of the bay's inhabitants. Beyond that, Kim Bassos and some of the others have continued to go out looking for the Boys whenever time, money, and weather allow, averaging maybe once every couple of months or so. Searches are easier, and therefore more frequent, in Misha's area than in Echo's.

They have found Misha regularly, still in his home range and accompanied by the usual cast of characters. He was seen seven times in 1992, six in 1993, and twice in 1994. One 1994 sighting was serendipitous—Kim and her husband were kayaking near Port Manatee when they spotted Misha and a couple of buddies deep-diving for fish. Pete paddled his kayak so fast that the dolphins came over and rode his bow-wave. One of Misha's companions, Left Bend Mid Nick, had been seen with Echo and Misha very soon after their release and now seems to have taken over for Davis as Misha's most frequent associate.

Echo didn't turn up again until a year after his last 1991 sighting, again during the annual fall survey of Tampa Bay. The survey team saw him twice during September of 1992, each time in Old Tampa Bay, not far from where he'd been seen the previous year. They found him twice more, still in the same area, during September of 1993. Each time Echo has been part of a larger group of dolphins, including several who appear to be regular associates. The number of sightings is so limited that we can't speak with any confidence, but so far his three most frequent companions have been dolphins called Screamer, Mimic 28, and Hoof and Bulge. Kim and others ventured up into Old Tampa Bay a couple times during 1994, but they didn't see Echo or any of his known buddies.

As of February 1995 we had accumulated a total of sixty-two sightings of Misha on fifty-nine different days and thirty-five sightings of Echo on thirty-two different days since their release. Kim Bassos recorded fortysome "focal-follow hours" from Echo and more than seventy from Misha. All our data and observations indicate that they have reacclimated every bit as well or even better than we could have hoped. Echo and Misha are again successfully leading the wild life.

So just what is "the wild life" in Echo and Misha's terms? What does the future hold in store for them? Day to day, their lives are probably relatively comfortable. I know any number of people, myself included, who have at times harbored fantasies of being reincarnated as wild dolphins. The life isn't necessarily as idyllic as it may first appear, however.

Dolphins have the same basic needs as any animal: food, protection, procreation. They also lead particularly active and complex social lives and sex lives.

The Sarasota and Tampa Bay dolphins eat a variety of fish, with mullet, pigfish, and grunt as the mainstay of their diet. They exhibit a variety of feeding techniques as well. In fact, bottlenose dolphins around the world are quite diverse in their feeding habits. They are generalists in many regards, much as we are. The differences are due in part to differences in habitat. But at least some techniques appear to be learned, not genetically encoded, and to be passed on to the young.

One of the most common techniques in the area is called "pinwheel feeding." The dolphin will swim around in a tight circle, trapping the fish in the center and forcing it to the surface, where the dolphin can then catch and eat the fish. This technique results in swirls at the surface of the water, one of the clues field researchers use to find and track the dolphins when they aren't visible at the surface.

Another favorite in the area is called "fish whacking." This is a shallow-water technique, often used on mullet. The dolphin may be swimming on its side, because in shallow water it can't get both an upstroke and a downstroke of its flukes when swimming upright. The mullet, trying to get away by using its greater maneuverability, turns 180 degrees back toward the dolphin's flukes. The dolphin then whacks the mullet with a powerful swipe of its flukes, sending it flying as high as thirty feet through the air. The fish is either killed or badly stunned in the process. The dolphin can then swim over and eat the fish at its leisure.

Kim Bassos reported seeing cooperative fish whacking, or dual fish whacks, in which two dolphins work together. One dolphin faces in one direction, one faces the other direction, and each kicks a fish out of the water. Then they each go get the fish their partner has whacked.

Perhaps there is competitive fish whacking as well. "Bet I can kick one higher than you can." It certainly looks like good sport.

Another method requires a seawall or similar barrier on one side. The dolphin does a lobtail of sorts, bringing its flukes down forcefully through the water, creating a sound that drives the fish toward the wall, where the dolphin can more readily snare it.

Bottlenose dolphins elsewhere devise their own techniques to suit their circumstances. In parts of Georgia, dolphins drive fish up the mud banks of rivers. The dolphin will slide up on the bank, totally out of the water, to gather its fish, then squirm its way back into the water.

Dolphins have little need for protection from the elements, and they don't build nests or lairs as so many birds and mammals do. One Sarasota dolphin, however, was consistently seen in a deep clearing in the middle of a shallow seagrass meadow. Because of this "nesting" tendency, he was one of the few in the area Randy and his crew never succeeded in catching. They named him Ace, since he was always in the hole.

A home range does appear to be vitally important, at least to the dolphins of Sarasota and Tampa bays. We didn't realize just how important until Echo returned to his core home range. Where an animal grows up, its physical adaptations to an area, environmental recognition, predator awareness, family, friends, and acquaintances probably all play a critical role in a dolphin's adjustment to life. Oceanic dolphins probably don't feel the same attraction to a particular area, and Tursiops in some parts of the world migrate seasonally.

Dolphins do need to safeguard against sharks and other life hazards. Instead of a nest or a lair, dolphins of all kinds turn to each other—the group or school or even just one good buddy—as their primary source of protection.

In the Sarasota bottlenose dolphin community, Randy has found that groups are organized on the basis of age, sex, family relationships, and reproductive status. Most adult females, along with their dependent calves, belong to one of several identifiable "female bands" in the area, each of which generally includes several unrelated matrilines. Daughters usually join their mother's band once they have their first calf.

These bands certainly do get together to socialize, mixing and matching in various combinations. The females interact with males in the area as well. But in general the members associate most consistently with others of their band. Within the band closest associations at a given time seem to be based on current reproductive status: calfless females tend to hang out together; moms tend to hang out with other moms who have kids of roughly the same age. Not altogether unlike humans.

Once a youngster is old enough to leave its mom, it joins a subadult group—the sort of groups Echo and Misha usually are found in these days. These groups tend to be the most active, with plenty of leaping, chasing, sexual behavior, and all manner of socializing. Within these larger, more fluid, subadult groups, each individual may have its most frequent associates, as Echo and Misha do.

Both males and females participate in subadult groups, though the ratio is skewed toward more males. Females tend to leave at a younger age, generally between six and ten, when they give birth and return to their mother's band. Males remain in subadult groups until the ages of twelve to fifteen, when they start associating less with the younger males and more with one or two other similar-aged males. These are the male pair bonds (or sometimes trios) Randy has referred to.

These pair bonds generally develop between males born into the same female band within a year or so of each other. It isn't a genetic bond but a social one. The bond seems to begin while the two youngsters are still with their mothers and then crystallizes later, as they mature, in the subadult group. This may happen even if the prospective pair joins the subadult group at different times. One calf, for example, stayed with his mom three years after his age-mate had left to join the subadult group, and the two of them still paired off later. Randy knows of at least one bonded pair whose younger brothers subsequently formed a pair bond of their own.

Such partnerships may last the rest of the animals' lives. Some males, lacking an appropriate age-mate, or perhaps for other reasons, never pair up. Whether a male forms such a bond or not may profoundly affect the lifestyle he will lead after leaving the subadult group. Paired males seem to travel more widely than do unbonded fellows. For example, at capture time in 1988 we saw Jimmy Durante and Norman, a Sarasota pair who have been together a dozen years or more, cruising through Tampa Bay, interacting with females. Single males tend to take a different strategy, staying closer to home near a female group.

Males who buddy up also tend to live longer than their solitary counterparts. Randy has recently started to look at survivorship and reproductive success with respect to social status in the Sarasota population. For females, those who raised their calves in the largest groups or the most stable groups had the greatest chance of seeing their offspring survive through the third year, the youngest age at which the kids are likely to leave home. Calves raised in small groups or in less stable groups are less likely to live that long. Hannah, the loner mom, clearly was trying to raise a calf under the most difficult of circumstances.

Among males, when they reach twenty years of age and older, those that are unpaired face a higher probability of dying during any given year than do those in pairs. Somewhat surprisingly, the remaining single males show a lower incidence of things like shark-bite scars than do members of the male pairs. Randy suspects this means that if a single male is attacked, he doesn't have anyone to watch out for him as he heals and therefore is likely to die of a wound from which a paired male might recover. Having a partner may not prevent all shark attacks, but at least the buddy can provide some level of vigilance and protection that allows the wounded animal to heal and acquire a scar.

By the time males are in their late teens, all the pairs that are going to develop have done so. Such a partnership, once formed, may be difficult, if not impossible, to replace. Randy has watched males whose buddies have died try out several other potential partners, but the pairings don't seem to last. Growing up together appears to be of critical importance in these pairings. A group of six males of roughly the same age, for example, were youngsters together in the largest of the Sarasota female bands. By 1980 they were all members of the same subadult group but then eventually split into three pairs. One of the pair members died several years later of pancreatitis. His buddy tried out a couple of potential partners, but none lasted. The survivor then functioned as a single male, spending most of his time on his own. Then, a couple years later, one of the other original six disappeared (presumed dead). His former buddy then paired up with the first lone male, forming what has become a solid, enduring bond.

Some dolphins in the area live to be in their fifties, but the vast majority die well before that age. As with humans, the females tend to outlive the males. The oldest known males in the Sarasota population are in their early forties.

The Sarasota dolphins seem to have several peaks in mortality. Many die during their first year of life. Calves of first-time mothers are especially prone to an early death. If they survive that year, youngsters typically have a good chance of making it until they leave their mother. Once they come through their initial year away from mom, they tend to fare well until they reach sexual maturity. Then a good many of the males die. As for humans, the teenage years can be stressful for dolphins; they are defining relationships and undergoing physiological changes, and males in particular may be more prone to getting themselves into dangerous situations.

Dolphins of the Sarasota and Tampa area face a variety of potential hazards, some of human origin, some natural. Excluding the mad slasher, anthropogenic causes include fishing nets, pollution, and boats (either disturbance from or collisions with). Dolphins and other marine mammals occasionally get tangled in nets and drown. It doesn't take anything that elaborate, however. Even the casually discarded plastic that holds a six-pack of beer together or a child's deflated balloon can have potentially lethal effects on an animal that becomes entangled in or swallows the foreign material. Holly took a series of photographs of a Hawaiian spinner dolphin who got caught in a floating loop of twine. The twine lodged itself over the animal's dorsal fin and under its belly. Slowly, over a period of a few weeks, the twine completely severed the dolphin's dorsal fin. (The researchers recording the process couldn't legally catch the animal to remove the twine without first obtaining a permit and it would have been nearly impossible to catch in the open sea, anyway.) The animal survived, but if the loop had been just a little tighter, it might have cut through the belly as well.

Natural hazards include diseases, predators, red tides, even fights with other dolphins. Viral infections have caused major die-offs along the east coast of the U.S. and in the Mediterranean in recent years. Sharks remain a threat, even to full-grown dolphins. Adults tend to fare better in a shark attack than do calves, since their blubber layer serves as a sort of armor that a shark has to bite through to get to vital tissues. This keeps wounds from being as severe as they might be otherwise. With calves it's far easier for a shark to do serious damage with a single chomp.

Males, as they reach maturity, engage in battles with other dolphins and physical competitions that can be deadly serious. Some males that end up in the stranding program have broken jawbones and serious wounds resulting from blunt trauma—probably from a jousting match with another male dolphin. These kinds of interactions probably come to a peak at about age twenty.

Randy has also described what he calls "border wars," seemingly aggressive bouts that occur at the boundary area between two home ranges. The warring parties are always large males, and the battles involve more than the usual aggression, in which everyone involved may end up bloody. The reason for such battles is unclear, since at many other times animals from one area may be in another's turf without hassle. These are relatively rare occurrences in the Sarasota area, Randy says—he sees it maybe twice a year. This sort of male-male aggression appears to be more prevalent, however, among the dolphins in Shark Bay, Australia.

Sexual behavior is a basic part of social intercourse among dolphins, occurring both between and within genders, across the ages, and throughout the year. Procreation is a somewhat different matter. Females come into estrus at certain times, generally late spring, summer, and, to a lesser extent, fall. Gestation is about a year, so most calves are born in late spring and early summer. Mating behavior is promiscuous—no monogamous pairings; no keeping of harems or the like, either.

Not every male gets to be a father. Preliminary genetic data on the Sarasota animals suggest that those males twenty years old or more are actually siring most of the calves. That's about the age the males reach full physical maturity—they're still growing until their late teens. Although they are sexually mature much earlier, they may not get to do anything productive (or reproductive) with it until they finish growing up. They may then be a more formidable opponent for other males and perhaps more appealing to the females.

Echo and Misha may have as much as thirty more years of life ahead of them, if all goes well. Food doesn't seem a problem. Both are in excellent body condition and seem to be having no trouble keeping themselves well fed. Barring a series of red tides or some environmental disaster that causes a major decline in fish populations, or serious overfishing in the area, both Echo's and Misha's home ranges should remain sufficiently bountiful to keep the Boys in fish for the rest of their lives. Randy and his crew have seen very few malnourished dolphins either in Sarasota or along the Tampa Bay shoreline, though some members of the offshore community in the deeper waters of Tampa Bay look skinny, with ribs showing.

Chances are good that both Echo and Misha will remain in their respective home ranges, now that they're back. Though Echo had become very much a part of the social grouping in Misha's area, the pull of home apparently was stronger. The Sarasota males don't seem to emigrate permanently, though they may make occasional forays into other regions.

The Boys' basic daily activities should remain pretty much as Kim Bassos has described them—traveling, feeding, socializing, playing, resting, and so forth—though the frequency of certain behaviors may change somewhat as they get older. For now, first thing in the morning they are likely to be traveling and feeding, generally in smaller groups of, say, three or four individuals. They will swim along, stop and eat, swim some more, stop and eat. Often the group will be widely dispersed, at some distance from one another, each dolphin feeding individually.

Then typically, come early afternoon, it's party time. One group may suddenly join another, and the two groups come together in a frenzy of leaping and chasing and carrying on. These gatherings may number ten to fifteen animals, mostly subadults, and can last anywhere from ten or fifteen minutes to an hour or two.

In late afternoon they might actually rest awhile, swimming more slowly, sedately. They may feed and socialize again in the evening, rest some more during the wee hours of the night, then start the early morning with some noisy interplay before getting serious about finding breakfast.

The Boys are at the age where their pairings, if any, should soon crystallize. They have reached sexual maturity—as of 1995 Misha is fourteen years old, Echo thirteen—and within the next couple years or so probably will be leaving the larger subadult groups and heading out with that special buddy or two—or on their own, if they don't have an appropriate age-mate with whom to bond.

Echo and Misha formed a bond with each other that clearly was important to them in their readjustment to the wild but apparently was not a lasting pair bond of the sort formed by some males who grow up together. It's possible they could still reunite and become a twosome again, but that now seems unlikely.

We don't know at this point whether either of them had a potential partner in his natal group, whether they are likely to lead a paired life, or go the single route. The membership of subadult groups tends to be very fluid—Echo thus far has been sighted at least once with ninety-two different dolphins, Misha with eighty. They each seem to have some regular companions, however. Misha may be in the process of pair-bonding with Left Bend Mid Nick. Then again, he has spent considerable time with Davis. Perhaps these three could even form a trio, since all of them have been together in the same group on many occasions. We've seen less of Echo since his return to Old Tampa Bay and know less about his potential partners. He does seem to have a number of possibilities, however. I confess, I especially like the notion of Echo matched with Screamer.

The Boys face some tough years ahead. They are at an age at which they may start taking trips out of their home ranges, heading into other regions to look for mating opportunities. That could put them in a position of being attacked by other males. They still have some growing to do and probably will have to wait another six years or more before they get the chance to reproduce.

We'd like to think they will ultimately be among the old-timers in their respective communities, producing some little Echo or Misha calves along the way. We hope to continue to keep track of them, checking in on them from time to time, to follow as much of their lives as possible.

Animals as sociable, as dependent upon one another, as dolphins need a good system of communication. Echo and Misha's post-release experiences—first beaching themselves twice, then fairly quickly acclimating and being accepted into the social structure, then ultimately splitting up, with Echo returning to his original home range—led Ken Norris to speculate that the dolphins in the area have a communications matrix going on under the water.

The Boys initially had trouble, Ken suggests, because they might have been "out of the loop." Then as they hooked into the matrix, they were able to get back into the swim, as it were. Finally Echo found his way home once he'd heard it through the grapevine.

This wider communication matrix, Ken contends, may hook all the dolphins in a piece of coast together. Not all at once—they don't necessarily need to broadcast along the whole length of the bay—but seriatim. Perhaps they have a longer-range communication that tells them where the sharks are. The dolphins may express a certain sort of agitation that is consistently associated with the arrival, in the area, of a dusky shark, for example.

Ken speculated that a system like that, spread out along the whole coast of Florida, might actually contain all the animals in it. An animal may know a lot about the world it's in as a result. Whatever wandering the dolphins do may still occur within that matrix.

This long-distance system says that if you're in a particular place, you are in the company of dolphins. And you probably know from experience who the dolphins are. You probably know fairly well where they are. You probably know something of their emotional state, as well—and perhaps a great deal more.

Unfortunately, at this point we simply don't know what they know, or how specific dolphins can be in their communication. I can't resist, however, briefly considering a question that has intrigued me all along: just what sort of "stories" might Echo and Misha themselves be telling about their adventures with humans? What, if anything, can they convey to their friends and relatives about where they've been and what they've been through?

Ken suspects these things would be absolutely uncommunicable. "I would think that their experiences in the world of humans are so strange by any measure in the whole history of dolphins, that it would have been like flying to Mars," he says. "They probably have no way to express that."

Randy points out (while noting that his evidence is purely anecdotal) that dolphins don't seem to be able to communicate even far less demanding messages. For example, during one of his capture-release projects, he was about to set the net around a group of dolphins that included several animals who had been caught earlier in the day. The previously caught dolphins split, apparently anticipating what was coming and deciding they'd already done their part. They did not, however, forewarn the remaining dolphins in the group, who stayed put and soon found themselves encircled by a net. Now, maybe the first set of animals was just in a "Let's save our own asses" mode. Or maybe that was an example of Tursiops humor—"Just a little practical joke, fellas." Or maybe, as Randy suggests, dolphins can't communicate in that sort of detail.

Kim Bassos, who spent the most time with Echo and Misha out in the wild watching their interactions with their buddies, says, "I definitely feel they had some way of communicating to these other dolphins something about us." She felt as though the Boys were bringing their friends over to check out the humans; at least other animals seemed to approach the boat and stare at the people much more often when one of the Boys, especially Misha, was part of the group. She couldn't help but wonder if he wasn't telling them, "There are those crazy people that made me do all those crazy things."

I'm inclined to agree, in part, with both views. I doubt that dolphins have the words, or other means, to converse about such matters. I suspect the specific nature of their experiences with us is, as Ken suggests, uncommunicable. That doesn't necessarily mean they can't convey something of their emotions, their feelings, about us—whatever exactly those might be. If I'm right about Josephine relaying to Arrow, in no uncertain terms, just how she felt about us that time she was so angry, then I imagine Echo and Misha could do likewise. I'll bet they can communicate something of their feelings about those strange creatures shrieking from the boat, some sense of their relationship with us—at least enough to pique the curiosity of their companions.

Even if Echo and Misha were able to recount the events of their past couple years in full, lurid detail, of course, there's no guarantee anyone would believe them: "Sure you were abducted by aliens. Tell me another one. . . ."

13

Alien Intelligence

So what does all this tell us about how smart dolphins are? If, as I have suggested, what we think of as intelligence relates to the degree of complexity and flexibility in an individual's behavior, I believe Echo and Misha have demonstrated considerable intellect. Their ability to adjust to life in captivity and then to readjust to life in the wild indicates considerable flexibility. Certainly in their life with humans the Boys were faced time and again with things their life back home never had taught them. Their reactions to their situation were complex, changing; they always found ways to surprise us. I think perhaps what I've found the most compelling evidence of their intelligence is their degree of individuality, their unique responses to life—their "dolphinalities."

Echo and Misha were about as closely matched a pair as you could get: they were the same gender and approximately the same age, both came from the same habitat and belonged to the same basic social system, they were accustomed to the same kinds of predators and the same kinds of foods, they were caught at the same time and underwent their captive experiences together. Yet they were two unique individuals, each with his distinct temperament and behavioral tendencies.

I suspect, if pressed, most of us who worked with the Boys would rate Echo as the "smarter" of the two. He was far more creative in the "innovative" tasks. He was quicker to pick up new tasks in training, and quicker to devise ways of getting out of them as well. After the first month or two of their stay, he generally came across as bolder and more decisive. He seemed more curious, to have more of an experimental sense, testing our reactions, say, to being nipped or being splashed. He seemed, in Holly's words, to understand the human condition better than Misha.

Still, I'm not convinced it's fair to label Echo as smarter than Misha; the two of them had different strengths and different inclinations, just as different people are better at different kinds of skills. I always had the sense that Misha was more at a loss in captivity than Echo. He was, I think, more a dolphin's dolphin. He seemed more ideally suited to the life from which he'd come. If I were a male dolphin in Tampa Bay and could choose a buddy to pair up with, I'd pick Misha in a flash, without hesitation. I'd bet on him as a steady, dependable, amiable, loyal, and sensible lifelong companion. With Misha at my side, I think I'd feel about as safe as any dolphin can feel amid the sharks and motorboats of Tampa Bay. If, on the other hand, I wanted help figuring something out, in some sort of problem-solving or brainstorming session, I'd pick Echo. If, as a human, I had to pick one for a show or research animal, again I'd pick Echo.

Part of what convinces me they are highly intelligent is this capacity for individual differences. This indicates that their behavior isn't all preprogrammed but must be learned, and learning, of whatever sort, is presumably what those big brains are for. Their brains are quite different from ours, however, so their intelligence probably is alien as well.

John Lilly's case for the superior intelligence of cetaceans, as well as for their language, "Delphinese," was based primarily on brain size. Some cetaceans do have extremely large brains, the largest on the planet; that much is certain. Just what that means is much less clear. Not all cetacean brains are so impressive, even in terms of absolute size. Delphinids—the dolphin family—are the truly brainy members of the clan. The sperm whale has the biggest brain, but it also has a much bigger body overall than the dolphins. Lilly dismisses body size as irrelevant, but he's about the only one to do so. In terms of what's called the "encephalization quotient," a measure of brain weight relative to body weight, Echo and Misha, as bottlenose dolphins, are the top of the line among delphinids, followed closely by Lags. They rank somewhat below humans but well above all other animals, including our closest relative, the chimpanzee.

You also have to consider the structure and organization of the brain. Human brains are notable for their great amount of neocortex, the outer layer and newest portion of the brain, which is generally associated with higher mental functions in us. Our brains show considerable convolution and infolding, which allows us to pack more cortex into a smaller area. Dolphin brains are even more convoluted than ours, giving them a much expanded cortical surface area. On the other hand, their cortex is thinner than ours, so in terms of total mass, their neocortex measures slightly less than a human's but well above all other comers, again including the chimp.

Cellular studies suggest that, on a finer scale, dolphin brains show a more primitive organization than do most modern mammals. Their brain cells are more uniform, less specialized, and certain layers of the cortex aren't as sharply defined. In these regards their brain structure resembles that of the earliest mammalian progenitors, of which the best living representative is the hedgehog. Dolphins' ancestors apparently broke off from the rest of the mammalian line to return to the sea before most of the sophisticated brain developments took place. Some researchers propose that dolphins' increased brain size may serve to compensate for this more primitive structure.

Just what that means in terms of their intelligence is debatable. I don't think it means that dolphins have all the smarts of a hedgehog. I do think it means that their brains—and their intelligence, however keen—may be even more alien than we might first imagine. Cetacean evolution followed a different tack from other mammals in so many regards, why not brain structure as well?

They are alien in other ways as well. The two halves of a dolphin's brain may function separately to a degree not found in other mammals. Not only do dolphins show independent eye movements, as mentioned earlier, but each brain hemisphere has its own blood supply, which isn't true of other mammals. Even more intriguing, dolphins may sleep with just one half of the brain at a time, with just one eye closed at a time.

Their breathing is conscious, voluntary; if they lose consciousness, they don't breathe and so they die—as early experimenters who tried putting dolphins under general anesthesia soon discovered. Among other things, their conscious breathing means they can't just settle in for a good night's sleep as we know it; they have to stay awake enough to keep breathing.

John Lilly first suggested that dolphins might get around this problem by sleeping with one half of the brain while the other half remains awake. Some electrophysiological data support the notion. A Russian researcher recorded brain waves in dolphins and described three stages of sleep; the stage of deepest sleep (stage three or slow-wave sleep) always occurred in just one hemisphere at a time, never in both simultaneously.

Dawn Goley's work with the Lags at Long Lab lends further support to the notion. She spent many a long night watching and videotaping the Lags and found that they consistently fell into a slow-moving "sleep formation," similar to what Ken Norris and his colleagues had described previously for spinner dolphins. The group would swim together silently, moving sedately and synchronously around the tank throughout much of the night. Dawn was able to add a new observation: the resting Lags kept one eye open, the other eye closed. The group periodically shifted positions, and everyone changed which eye was open, which closed. At first Dawn expected that they would each keep an eye open to the outside world, watching where they were going, and close the eye toward their schoolmates. Just the opposite; they always kept an eye out for each other. That makes sense when you think about sleeping in the open ocean: you aren't likely to run into anything out there; but if you lose track of your school, you're in the deepest possible trouble.

Echo and Misha, though they rested on the bottom of the tank rather than swimming in formation, seemed also to close alternate eyes as they rested. We couldn't tell for sure, however, since they rested at some distance from the windows, their eyes hidden in the darkness, whereas the Lags passed close to the window each time they circled the tank.

Dolphins appear to sleep, after a fashion, but they may not dream at all. In his many hours of recording sleep waves, the Russian scientist never observed any rapid eye movement sleep (REM), which in humans and other mammals is associated with dreaming.

Based on brain considerations alone, there's good reason to believe dolphins are smarter than your average animal. Brains are metabolically expensive, and evolution isn't going to favor a bigger one unless it somehow more than pays for its upkeep. We still don't know just what dolphins are doing with those large brains, however.

Researchers have offered a variety of suggestions, besides the notion that increased size compensates for primitive structure. Another possibility is that the conscious, voluntary control of such bodily functions as breathing, which for us and most other mammals are handled unconsciously, requires more brain power. This seems unlikely, at least as the primary explanation, since all cetaceans share this trait, even those with relatively small brains. It doesn't tell us why, for example, a Lag might have a brain twice the size of a river dolphin of comparable body size. The need to keep the brain supplied with oxygen during dives has also been offered in explanation, but it meets the same problem as the conscious-breathing notion: it's true of all cetaceans, regardless of brain size; it's true as well of seals and sea lions, whose brains are considerably smaller than a dolphin's.

Their unusual sleep patterns and apparent lack of dreams (or lack of REM, at least) has also been implicated. If dolphins do indeed sleep with only one half the brain at a time, and the two brain hemispheres are functionally separate to some degree, perhaps the large size is at least partly a matter of duplication—they carry a spare brain so they can keep one going while they shut down the other one for sleep. "The creature with two brains"—it's aliens again.

Dreamless sleep also could be a factor. Francis Crick, the Nobel Prize–winning codiscoverer of the structure of DNA, has proposed that the purpose of REM sleep may be to create additional space in the brain by clearing out old memories to make way for new ones, a process he calls "reverse learning." If dolphins can't use dreams to do this, they may need some other means of preventing neural overload—such as a particularly large brain. However, this is only one of many interpretations of dreaming, much less of dolphin brains.

I consider two potential reasons for a big brain the most likely candidates: the extensive development of the auditory system in support of echolocation, and the complexity of their social lives and communication skills. Dolphins' auditory systems are clearly much expanded over those of other animals, and they probably devote a considerable amount of brain space to processing echolocation. A large brain isn't necessarily required for echolocation—bats manage quite nicely with much smaller brains. But dolphins not only acquire more information from their echolocation than do bats (regarding the internal structure of an object, for example), they may also engage in more fancy brain work based on what they find, beyond just the basic navigation and prey detection.

Dolphins also engage in complex social relationships requiring sophisticated communication and considerable learning. This is true as well of other large-brained mammals—humans, chimpanzees and other higher primates, elephants, and the large social carnivores, such as wolves. I suspect sociality was the major incentive for the development of big brains in all these animals. These complex social systems generally involve pronounced division of labor, interdependence, and group cohesion and permanence. They also tend to be characterized by long life spans, low reproductive rates, small litter size, a communication system that is both subtle and expressive, and an extended "childhood" period devoted to learning the rules of the society.

We humans are distinguished by our especially prolonged youth, with continued brain growth throughout this period. We are born with brains only about a quarter of their adult weight, which then take seventeen or eighteen years to reach full development. Other mammals are born with brains much closer to their final adult size and, presumably, a smaller capacity for learning and development. In some lower primates, for example, the brain at birth is already 80 percent of adult size.

Dolphins are born with brains at a more advanced stage of development than ours—just over 40 percent their ultimate weight.

They need a longer gestation period, about twelve month in Tursiops, so the newborn can swim, breathe voluntarily, and so on from
the outset, which undoubtedly requires a more developed brain.
Bottlenose dolphins apparently reach full brain development at age
nine or ten, about half the total growth period required for human
brains but considerably longer than for most higher mammals. This
suggests they've got much learning to do.

Part of the answer to the question of intelligence, for dolphins as for
any animal—ourselves included—is that they're basically as smart
as they need to be to make their living. The question then becomes:
what does their lifestyle require of them? What do they need to be
able to perceive, to learn, to understand, to remember, to communicate, to create?

Dolphins, like our early ancestors, are both prey animals and
predators. They must protect themselves against attacks by sharks,
as well as find and catch fast-moving fish for their suppers. Dolphins' echolocation evolved to meet these two needs; so did their
sociality. These are the two principal incentives for social living: the
increased surveillance and protection provided by the group and the
opportunity for cooperative food-finding efforts. Young dolphins
have a great deal to learn not only about how to find their way
around their physical environment but also about how to function
within their social milieu.

Just what you need to know to be a part of the group varies
across species and even within species, exhibiting what could be
called cultural differences among groups. Ken defines culture as
"information transferred by teaching or observational learning."
Dolphins clearly transfer a great deal of information in this fashion.
For example, Echo and Misha may employ such feeding strategies
as "fish whacking," whereas bottlenose dolphins in other parts of
the world will chase fish up onto mud banks, or even work coopera-

tively with human fishermen. These techniques seem to be learned and passed on to the next generation. Apparent differences between the bottlenose dolphins of Sarasota Bay and those of Shark Bay, Australia, in the degree of male-male aggression and in the tendency of males to "gang up" on fertile females also may represent learned, "cultural" differences. While such behavioral variation may well be the product of ecological variation, the same might be said of much human cultural diversity as well. One sociobiologist has suggested, for example, that peoples living near the equator tend to eat hotter, spicier foods than those of us in more temperate zones; intestinal parasites are much more common in the tropics, he argues, and hot chili peppers and the like may help discourage parasites.

Social living also requires a keen sensitivity to the emotions and reactions of your companions. Every dolphin trainer or handler I've talked to seems to agree that the matters most readily communicated between humans and dolphins are basic emotions. The dolphins may be even better at picking up on our emotions than we are on theirs. As Holly suggested, at least in some regards, Echo and Misha may have understood us better than we understood them. That advantage may stem from a lack of language per se. We can lie to or mislead each other—and ourselves—with words, saying we're feeling one thing when we're actually feeling something quite different. Dolphins (and other animals) aren't fooled by our words; they simply respond to whatever cues, subtle or otherwise, we may give out.

I suspect that may be behind what I've sometimes heard described as mental telepathy in dolphins. They are highly social creatures. Indeed, their survival may often depend on their ability to stay attuned to their companions. In a captive situation they are likely to turn their considerable social savvy toward humans.

Around the turn of the century a horse named Hans was all the rage in Germany for his apparent mathematical skills. He accurately answered a wide range of arithmetic problems by tapping out the correct number with his hoof. Ultimately, some discerning scientist noticed that, rather than adding and subtracting, Hans was responding to very subtle cues from his trainer. The trainer jerked his head slightly when Hans reached the appropriate number of taps; all the horse had to do was keep tapping until he noticed the head jerk. If the trainer didn't know the correct response, the horse got it wrong, too. It wasn't a deliberate hoax, apparently—the trainer really thought the horse could add and subtract.

The example of "Clever Hans" has been cited repeatedly in animal intelligence debates ever since, generally to disparage the accomplishments of nonhuman species. Many detractors overlook just how clever Hans really was. Okay, maybe he couldn't add. But he could detect some very subtle cues that most of the people around him were missing—even the person giving the cues.

I suspect many animals can read a great deal in our behavior, not just emotions but also small signals conveying what we'd like them to do. Dolphins, as highly social creatures, may be especially adept at this. They probably get even better at it the more they learn of human ways, and the more they get to know a given human. What could seem like telepathy—a dolphin performing a particular trick the trainer had in mind but hadn't overtly signaled, for example—may often result from a clever dolphin cuing in on inadvertent signals from the trainer. We don't necessarily need to revert to parapsychic explanations for this sort of dolphin behavior, but that doesn't make it any less impressive, either.

Dolphins' social nature also seems to predispose them to cooperative behavior, to assisting one another in various ways. They may stand by a fellow dolphin in distress, swim around an injured animal, charge its attacker, or actively offer support such as pushing a sick animal to the surface for air. Josephine, for example, stayed by Arrow's side as she was dying, stroking her and helping keep her at the surface. Dolphins even come to the aid of other species of cetaceans. There have also been reports, dating back to ancient Greece, of dolphins rescuing drowning humans, holding them at the surface or even pushing them to shore. I know of no documented cases of such rescues. Given their tendency to hold ailing animals to the surface to breathe, however, I could easily see a dolphin doing the same for a human. I'm less sure about which way they'd push—while surfaceward versus downward makes considerable difference to a dolphin, shoreward versus out to sea isn't likely to mean the same thing to them as it does to us. As the joke goes, "Nobody talks about how many people the dolphins push back out to sea."

Some critics argue that this helping behavior is purely instinctive; the dolphins don't "know" what they are doing when they come to the aid of another. That seems unlikely; given their capacity for learned behavior, the behavior probably is at least partially conscious and volitional. This behavior may be more often directed toward relatives or close associates, but it is by no means restricted to that. Such caregiving behavior may be an example of what's called "reciprocal altruism." Dolphins are likely to behave in altruistic ways because sociality is such a basic part of their lives and they are likely to need similar help at some point themselves—you're more likely to get help when you need it if you give it to others of your group when they need it. They seem to extend the principle beyond their immediate group, even beyond species boundaries.

Group living also means that certain behaviors simply can't be allowed, for survival of the group as a whole. Some of these are surely taught as "rules" to the young. They also seem to apply some sense of "etiquette" or "ethics" toward humans. Anyone who's worked with dolphins probably has experienced incidents in which the animal clearly seemed to behave more cooperatively or more gently than expected. Echo and Misha, for all their nipping at our fingers from time to time, never seriously tried to injure any of us, though I suspect there may have been times they felt like it.

Once when I was swimming with Jo and Arrow at the Dolphin Research Center, I believe Jo decided after a while that she'd had enough of my presence and wanted me out. In retrospect, in my eagerness to swim with her, I think I ignored several previous signals she gave me. Finally she swam over to me and thwacked me on the leg with her flukes. It was enough to let me know I'd been hit and to leave something of a bruise, but nothing compared to what she was capable of inflicting. I took the hint and got out. I was left with the sense, though, that she'd made a very deliberate choice regarding the amount of force necessary to make her point—and, I think, had tried nonviolent means first.

Schooling behavior of the kind found among oceanic species, such as Lags and spinner dolphins, adds yet another twist to dolphin sociality, bringing with it some loss in individuality. The school serves as a form of protection against predators, but schooling behavior (like flocking or herding) works as a defense only if everyone in the group essentially looks the same and acts the same. Schooling dolphins tend to show relatively little in the way of individual differences in physical appearance. They all tend to swim continuously in the same general direction at the same general speed. This fools the eye of a predator in what's known as the "confusion effect." When every member of the school or flock looks alike, the predator doesn't have a clear-cut reference point to focus on and so can't coordinate an attack properly. Researchers found that if they dyed one fish blue in a school of silver fish, not only was the blue fish more likely to get caught, but so were the silver ones right around it; the color difference gave the predator a reference point.

"Because when under attack," Ken Norris says, "whether they be spinner dolphins or anchovies, each school member must conform to iron rules, individuality is reduced close to zero." The rigid schooling behavior drops away to a large extent when they aren't under attack, allowing for greater individuality, but the conformism may spill over into the rest of their lives. Schooling dolphins are cooperators, utterly alone outside their schools, which may be why they are generally so docile upon capture and in captivity.

This loss of individuality seems alien, even a bit terrifying, especially to us in Western culture. On the other hand, schooling may offer dolphins an experience we can't fully imagine, making them part of a larger unit, a larger self. Their senses become integrated with those of their schoolmates, all of them functioning in unison like members of a chorus line and forming a sensory system that is both faster and more potent than any individual could manage. Ken calls this the school's "sensory integration system."

Coastal bottlenose dolphins, such as Echo and Misha, are highly sociable, but they aren't true schooling animals. Here the groups are smaller, and the protection provided by having companions around isn't a function of fooling the predator but of increased vigilance provided by the additional eyes and ears. Here there is room for a considerably more individualistic approach to life.

Again, Echo and Misha are unique individuals, each with his own strengths and proclivities. Species likewise differ in their special abilities. Yet we tend to gauge intelligence largely based on language and manual skills. How do you apply that to an animal that lacks both language and hands?

Psychologist Howard Gardner introduced the notion of "multiple intelligences" in people, arguing that you can't measure intelligence in terms of a single IQ score. He describes seven "candidate" intelligences: linguistic, logical-mathematical, musical, spatial, bodily-kinesthetic (movement), interpersonal, and intrapersonal. He certainly had no intention of applying these to nonhuman animals; they aren't meant even to rate one human against another. Still, they provide some framework for talking about the different kinds of capabilities various species may have.

Humans undoubtedly are best both at language and at math and logic skills. That's what gives us our advantage, such as it is, over other animals. We're probably best, too, at music—at least music in our terms. But other animals may equal or better us on other sorts of intelligences, whether the ones Gardner described or kinds we generally don't consider.

I'd argue that all seven of Gardner's human intelligences can be found, in some form and to some degree, in other animals. Dolphins show elements of all of them. As Lou Herman's work has illustrated, they can learn some aspects of human language, though chimps thus far have proved more impressive in that regard. If you interpret this intelligence more broadly as communication skills, however, dolphins probably are at least on a par with chimps and other highly social mammals. They have a large and varied repertoire of vocalizations, and their complex social system gives them plenty to communicate about.

Although various animals have shown some counting ability in the sense of somehow being able to keep track of the number of items or events, math probably is not a strong suit for nonhuman animals (nor for many humans, for that matter). There are undoubtedly many instantaneous computations involved in dolphin echolocation; whether that would rate as mathematical as such I'm not sure. Logic, if considered as basic problem solving, is undoubtedly more prevalent in the animal kingdom. (Our Siberian husky puppy has figured his way out of every enclosure we've devised for him thus far.) Dolphins have repeatedly demonstrated their cleverness in this regard, both in captive situations and in the wild. A bottlenose dolphin at Marine World, for example, had been trained to pick up stray bits of refuse that landed in the pool and bring them to a trainer in exchange for a fish reward. One day, apparently, the dolphin snagged a large piece of newspaper. Rather than turning the whole thing in for one fish, the dolphin stashed the paper in a corner of the pool and turned it in a small piece at a time, milking the trainers for all the fish he could get.

Certain birds have definite musical inclinations. Not only do many of them sing songs of their own, some have proved able in laboratory studies to discriminate among different rhythms or melody patterns. The complex songs of the humpback whale are well known. A sperm whale once established a duet with a submarine's Fathometer, weaving an intricate rhythm of clicks around the regular "pings" produced by the sub.

Dolphins' responsiveness to music has been noted—albeit largely in anecdotal form—at least since the time of Pliny the Younger (circa A.D. 100), who wrote, "The dolphin carrieth a loving affection not only unto man, but also to music; delighted is he with harmony in song, but especially with the sound of the water instrument or such kinds of pipes." I've seen a videotape of several captive dolphins moving their bodies in what looked for all the world like a dance in response to a live performance by Kenny Loggins.

None of this compares to a Bach or a Mozart, perhaps. But dolphins, among other animals, do seem to have at least some appreciation of human music and most certainly have a good sense of rhythm in other contexts. Dolphin echolocation involves very complex patternings of sound—perhaps as much or more so than either human language or music—and our research with Echo and Misha suggests that they may use a strong sense of rhythm to help them sort it out.

With respect to both spatial and body-movement intelligences, I suspect that various other animals, including dolphins, may equal if not surpass us. Dolphins live within a more fully three-dimensional space than we do, since they can move up and down in the water column, whereas we function largely on a single plane. The same is true for birds and arboreal primates. Their internal representations of space, just for day-to-day living, must be expanded accordingly. Dolphin echolocation, while sound based, gives the animals graphic spatial representations of the world around them.

Many animals, most certainly dolphins among them, are capable of graceful, intricate movement patterns, showing considerable awareness and control of their bodies and how they move. Anyone who's seen a dolphin show knows something of the finely coordinated leaps they're capable of—often with several animals working in synchrony. Spinner dolphins may do the most spectacular aerial maneuvers of all, executing amazing spiraling leaps. Bow-riding wild dolphins also display grace, coordination, and great precision as they weave in and out within inches of the bow of a moving boat. While leaping may serve various functional purposes, and bow-riding may be a way of hitching a ride, they also appear to be great fun. There seems to be an aesthetic involved, perhaps even some competition among friends.

Gardner's last two intelligences are what he calls the "personal" intelligences, which by definition excludes other animals. "Interpersonal" intelligence—ability to deal with other people—could be redefined as something like social skills to apply to a much broader range of species. Social skills may be every bit as important and as complex for dolphins (as well as for chimps, elephants, social carnivores) as they are for us. Dolphins need to keep track of and interact with perhaps hundreds, even thousands, of their companions. They may have a smaller number of regular associates, with whom they must cement and maintain bonds, as well as more fleeting, though perhaps frequent, interactions with many more. The extent and nature of the relationships will vary with species and habitat. Males may form intimate, long-lasting pair bonds or shorter-term coalitions to sequester females for mating. There are apparent friendships and fights, disagreements and making up. The school is of utmost importance for most dolphins, and they do seem to have to learn their manners as they grow and develop the appropriate social graces to survive. You may see the occasional loner or apparent outcast, at least among coastal bottlenose, but that's a rarity—and probably unheard of, a death warrant even, among oceanic species.

"Intrapersonal" intelligence—knowing oneself—is tougher to address in other species. We don't know them well enough to get a sense of how well they may know themselves. While I suspect most animals, if not all, have a sense of self versus nonself, a sense of self-awareness or self-consciousness may be more limited, both in scope and in frequency of occurrence. Our language probably predisposes us to a kind of self-reflection not possible in the less linguistically inclined.

Chimpanzees can recognize themselves in a mirror, which is considered an indication of self-awareness. When researchers place a dark smudge on the animal's forehead while it's sleeping, as soon as it wakes up and sees itself in a mirror, the chimp rubs it off. Most other animals react to their mirror image as if it were another animal. Similar mirror tests with dolphins thus far have been suggestive but inconclusive. Part of the problem is that dolphins don't sleep in the normal fashion, so the researchers have to smear the paint on while the animal is awake and aware. Several dolphins in one study did indeed posture and turn in front of a mirror placed in their tank, as if checking themselves out; others, however, showed no apparent reaction to the smudges. The mirror test might be more appropriate for humans and other higher primates, for whom facial recognition is important, than for dolphins. Given their keen responsiveness to sound, perhaps some auditory version of the test might be more telling. I'd argue, in fact, that dolphin signature whistles—their individual "names"—indicate at least a degree of self-awareness.

If we rank animal intelligence from an anthropocentric viewpoint—how do they compare with us?—of course we always come out on top. Even in these terms, however, I think dolphins rank among the leading contenders.

If another species were devising the list, the important intelligences might look very different, and we might not fare nearly so well. I believe echolocation, for example, would meet Gardner's criteria for an intelligence.

If echolocation were to be considered an intelligence, dolphins would surely stand at the head of the class (actually, beluga whales probably would come first). Bats would also score well, though perhaps not quite as well as dolphins. Oilbirds and cave swiftlets, plus a few rodents, have a primitive echolocation ability—better than ours but nothing compared to dolphins. We have some capability in that regard but it's very limited, not counting our sonar equipment—which still is no match for a dolphin. Blind people can learn to get around in part by tapping a cane and listening to the returning echoes. But I suspect the most skilled human listener's echolocation intelligence probably doesn't even compare to the linguistic intelligence of those dolphins trained in human language.

14

Dolphins Mortal,
Dolphins Divine

Diviner than the dolphin is nothing yet created;
for indeed they were aforetime men and lived in cities along
with mortals, but by the devising of Dionysus they exchanged the
land for the sea and put on the form of fishes.
—OPPIAN

During the press conference the day before Echo and Misha's transport back to Florida one reporter hit us with a zinger: "Does what you learned justify kidnapping the dolphins for two years?" I couldn't speak. Howard took over. I've forgotten what he said. I think I may have added something—I've forgotten that, too. Whatever we said wasn't adequate, wasn't satisfying. I later heard the radio piece by the reporter who'd asked that question; he made no mention of the "kidnapping" question, and it turned out to be one of the fairest, most informative reports of the lot.

But the question stuck with me. It was a major impetus for writing this book. This book is the answer I couldn't give then, the response that wouldn't fit into a sound bite.

First, I resent the word "kidnapping," which implies illegal deeds. Perhaps a dolphin tribunal might justly so accuse us. In human terms, however, our capture of Echo and Misha was entirely legal—we went to great lengths to get all the necessary permits and to care for the dolphins appropriately. I'd liken it more to a military draft than to either a kidnapping, which makes it sound criminal, or a sabbatical, which makes it sound voluntary.

I don't much like the word "justify," either. That seems to imply some tally or accounting system—how much data, how much knowledge, does it take to balance out two years of the dolphins' lives? Make that two years that are *different* from what they otherwise would have been; it doesn't necessarily mean the years were lost or even bad. Only Echo and Misha could tell us that.

My simple answer is "Yes, I think our project was worthwhile." Of course, I find that easier to say now that I know the whole thing turned out well; if the Boys hadn't readapted so beautifully, if something had gone wrong, my response undoubtedly would be more qualified. I'd love to hear Echo's and Misha's answers, but all we know on that score is that the experience doesn't seem to have harmed them or kept them from pursuing normal dolphin lives. I'd like to think it enriched their lives in some way, broadened their horizons, gave them a perspective the other "ordinary dolphins" out there simply couldn't have.

I've tried in this book to share some portion of what we learned—not so much the scientific data as the whole experience— and thus allow readers to answer the question for themselves. I decided to tell Echo and Misha's story in hopes of extending the impact of what we learned even further; the more people who learn something about dolphins, their world, and their relationship to us, the more valuable Echo and Misha's venture into our world becomes.

I don't think we'll ever really understand dolphins without a better understanding of their echolocation, and I believe the Boys made an important contribution in this regard. Studying echolocation among wild dolphins remains exceedingly difficult, but we tried to take a step toward more natural conditions by looking at free-swimming animals. The more we can learn of their collateral behaviors—what they *do* as they echolocate—and how these movement patterns relate to the sound patterns, the better equipped we should be to study them in increasingly natural situations, and even in the wild.

Among the many things we learned from the dolphin sabbatical program is that the "sabbatical" process itself probably isn't such a good idea, as Ken himself concluded. To bring animals in for a short period and then release them puts them under maximum strain—Echo and Misha spent much of their time adjusting, first to captivity and then again to the wild. These adjustments may be especially stressful if the animals need extensive, complex training, whether for research or show purposes. Better, perhaps, to have a few dolphins adapt to captivity and remain there than to have a number of animals coming and going all the time.

One situation in which the sabbatical notion might be useful is if we need to expand the gene pool among the breeding populations of captive dolphins: a couple of male dolphins, perhaps a known pair bond, could be brought into captivity briefly for breeding purposes. The period could be considerably shorter than our two years with Echo and Misha. Little training would be involved, other than teaching them to eat dead fish and a few other necessities, further easing their transitions. If the breeding colony were in bay pens near the home range of the newly caught dolphins, adjustment and readjustment should be even easier. Current numbers of captive dolphins appear to be sufficient, however, to maintain genetic diversity without resorting to even these short-term captures.

Echo and Misha's "sabbatical" program helped establish some important guidelines for future reintroduction attempts, principles that weren't followed in most prior releases. According to the National Marine Fisheries Service, more than seventy dolphins in the U.S. have been released since 1964, after periods ranging from a few days to ten years in captivity. Very little information is available on most of these releases, however, and none of them conducted a systematic follow-up to determine how well the released dolphins readapted to life in the wild.

Three of the more recent release projects offer some documentation of their efforts. In two cases the dolphins were released at a considerable distance from their original capture sites; in fact, one Pacific bottlenose caught near Taiwan was released in the Atlantic. In one case no attempt had been made ahead of time to evaluate the dolphin population or the food supply in the area. Several of the animals had been in captivity a dozen years or more. An adult male and an adult female released together—not a natural pairing—separated almost immediately. Reports of the dolphins following release came mostly from local fishermen who were untrained in assessing body condition or behavior. One release in Australia involved five adult wild-caught dolphins, plus four captive-born calves, including one newborn. The newborn died or disappeared, and its mother was seen begging from boats. Several appeared to lose significant amounts of weight; three were ultimately recaptured.

Echo and Misha's reintroduction to the wild was unique in several regards, including its success and the fact that we'd planned the release even before the capture. Not all captive dolphins can or should be released. Arrow, for example, simply wasn't healthy enough to be a good candidate for release. If Misha hadn't recovered fully from his illness, we wouldn't have released him, either. We monitored his health every step of the way and were prepared to postpone or call off his release at any point if need be. Josephine may have been in captivity too long to make the transition back to the wild either comfortable or desirable for her. We ultimately decided against trying to release her because she seemed perfectly happy staying at the DRC. In fact, one time there was a hole in the pen Jo was in, and several other dolphins swam out temporarily, though they neither ventured very far nor stayed out for very long. Jo didn't even budge.

The one thing we wished we could have done differently in Echo and Misha's release was to have an acclimation pen right at the release site. If we'd been able to use bay pens back at Ruskin as we'd originally planned, we could have just opened the pen on release day and let them proceed from there, at their own pace, rather than having to haul them around in a boat for two hours before plopping them down in an area Misha hadn't seen for more than two years, Echo maybe never. In retrospect, even with the halfway house at Mote, it might have helped if we'd held them in a makeshift pen, just for a few hours maybe, at the release spot itself to give them time to get oriented before we opened the gates and sent them on their way. That might have eased the apparent disorientation that led them to beach themselves.

Based on our experiences with Echo and Misha, future releases should consider at least the following: (1) the age and sex of the dolphin, with younger animals being better candidates, and natural social groupings (such as male pair bonds) most desirable; (2) releasing the animal near its original home range; (3) time spent in captivity—the shorter the better; (4) establishing a "halfway house" pen at or near the release site; (5) conducting surveys of the local dolphin population before the release and assessing the food supply in the area; and (6) conducting a systematic follow-up study for at least a year following release.

The principles we learned from Echo and Misha's successful reintroduction to the wild are already finding broader application. The National Marine Fisheries Service now insists that reintroduction programs include the sort of follow-up monitoring we carried

out with the Boys. What we learned—thanks especially to Echo—
about the importance of taking them home again has recently bene-
fited a couple of stranded dolphins. Both animals were rehabilitated
at Mote Lab and then returned to their home ranges, after 37 days
in one case and 107 days in the other. The latter dolphin, called
Freeway because a driver had found him on a bank near the inter-
state, came from Misha's area. He'd suffered pneumonia, a collapsed lung, and serious shark-bite wounds. The other animal had
been caught in a crab trapline. Both were monitored following
release and are doing fine.

I believe that all of us, especially those most intimately involved in
the project—including Ken Norris—learned from our experiences
with the Boys and gained personally from them. I don't know if
that counts in our "justification" tally, or if it's purely selfish, but I
think it's important, regardless.

Echo and Misha enriched all our lives. We learned something
about dolphins, about training, about ourselves, and about each
other, as well as getting to know two very special individuals. The
group, particularly the early team, was a close-knit one, and now
that everyone has scattered across the country, some still refer long-
ingly to our time together as Camelot.

Ken told me later he'd been bowled over by the emotional
reactions of the crew, by the degree of our engagement with the
dolphins and the work we were doing with them. Back in the early
days in the business, when researchers would shoot dolphins to
collect their bones for museums, he'd learned to push much of the
emotional content aside. Faced with our sometimes tempestuous
feelings about what we were doing, he opened himself up in a way
he hadn't before. "Like so many times in my career," he says, "I
was pulled by the suspenders into the present day by my younger
friends."

As for me, I learned a whole range of things. That was another major impetus for writing this book—the experience meant so much to me, over and above the scientific research, that just writing a doctoral dissertation didn't feel nearly adequate to express it. I needed to tell the whole story, in part to get a clearer view of the larger picture myself, to understand the echolocation work better by placing it within a broader framework.

I didn't learn to "talk with the dolphins" in the fashion I'd once hoped, but I did learn to not be disappointed by that. Echo and Misha are far more interesting, I think, when you get to know them as unique individuals rather than mythologizing them, looking to them for our own private wish fulfillment. Somewhere in the process I think I even learned to talk better with people.

While I definitely think our project was worthwhile, that doesn't mean I'd do it again. Nor, I suspect, would most of the rest of the crew, though everyone I've talked to since is glad to have been a part of it. Most of us have at least somewhat mixed feelings regarding keeping dolphins in captivity; some would be unwilling to work with captive dolphins again.

For my part, I believe in the value of and the need for keeping some dolphins in captivity. People need to get to know and learn to care about these marvelous creatures—not just dolphins but the many other co-inhabitants of this planet as well. Direct contact is vital to that; a moment of eye contact, a brief touch, or a game of Frisbee can be more powerful than all the shows and demonstrations.

Echo and Misha touched a great many lives during their two years in captivity, and not just those of us on the dolphin crew. Long Lab is open to the public, and thousands of people come through on tours each year. If even just a few of those who saw the Boys came away with greater respect and concern for dolphins, that can make a difference. I doubt there would have been the public outcry over dolphins killed in tuna nets if so many people hadn't had the opportunity to appreciate dolphins in zoos and aquariums.

Animal-rights activists may argue that we should go out and see them in the wild. Most people never get that opportunity. If they did, if we could arrange for all the millions of people who see captive dolphins each year to get out to see dolphins in their natural habitat, we would undoubtedly destroy that habitat in the process, inundating the dolphins with our presence. The boat traffic in Sarasota and Tampa bays is hazardous enough for dolphins as it is (and even more so for the slow-moving manatees); the place would be shore-to-shore boats if all the busloads of people who see dolphins in zoos and aquariums each day decided to go there instead. Better, I think, to have a few dolphins in captivity as representatives, emissaries from their kind to ours.

Some will argue that we should simply leave the dolphins, and all God's creatures, in peace. I don't think we can do that. We need to face the fact that we *are* powerful creatures and we *do* have an impact on the other inhabitants of this planet, whether we like it or not. We can't maintain a "hands off" stance, because our hands are everywhere already. At least for the time being, our best bet, I believe, is to get to know our many "sharers-of-the-Earth," as Ken puts it, so we can learn to make our impact as positive and as gentle as possible.

If we are to keep dolphins in captivity, it needs to be done well. I think we should have relatively few dolphin facilities—I'm definitely not in favor of letting every hotel have a couple dolphins in a tank to entertain the guests—but make those few especially good ones. As we learn more about dolphins, we can provide better facilities for them. For one thing, tanks need to be larger than the current requirements specify. I never felt the Boys really had room enough to swim and jump and play freely at Long Lab, especially in the thirty-foot pool.

Animal-rightists may argue that no tank is big enough. True, we have no way of knowing what amount of space Echo and Misha would consider adequate. The tank should be large enough, though, to give the dolphins not only room to exercise properly, but space enough to get away from each other and from the humans should they wish. At least as important as size is the matter of making their environment interesting enough for them.

One key to making the environment stimulating is to have others of their own kind in sufficient numbers, preferably enough in a given facility so that the animals can have something resembling a normal social life. Having just two dolphins in a tank, as we did with the Boys and as many other facilities have done, simply isn't adequate. Part of the reason Echo and Misha tended to be quieter and less active than the Lags, I think, was that there were only two of them.

Dolphins are highly social creatures, and many species seem to live in fission-fusion societies, where small groups come together into larger groups, and where group composition may change from day to day. Like us, they may have their regular "best friends," but they don't hang out with just one individual all the time. Maybe they don't feel like dealing with somebody on a given day; they should have more than one option. Captive dolphins need to be able to carry on a full and active social life, even without considering the humans as part of their social system. Having just two dolphins may make the animals that much more inclined toward interaction with humans, since that's the only social variation they can get—which could be an advantage in training. But I think it's probably a bigger advantage to have a lively, active dolphin to work with.

Ideally, the physical environment should be made interesting as well. That can be hard to do in a concrete tank, where it will take greater effort to make the setup stimulating for the animals, with toys and various things to do available for them. "Behavioral enrichment" for captive animals of various sorts is an important trend these days. For dolphins this can be somewhat easier in bay pens, since they will have all the natural sights and sounds and tastes of the ocean, often complete with both plant and animal life.

Part of the picture, too, is to have experienced trainers and handlers who care deeply about the animals and are interested in them as the unique individuals they are, plus a program that respects the animals and teaches the public to do likewise.

I think there are facilities around that manage to do these things well. We can't know for sure just how the dolphins feel about the matter, but there are occasions when they can vote with their flukes, as it were. For example, at the Dolphin Research Center, where Josephine now resides, the dolphins can—and sometimes do—get out of the pens, but they choose to stick around. When Hurricane Andrew passed through Florida in 1992, the DRC opened up all its pens to ensure that the animals wouldn't get caught in the netting. All but one returned.

Other facilities can report the same, where the captive dolphins, either accidentally or deliberately, are given access to their freedom and choose time and again not to take it. The Navy does open ocean work with some of its dolphins—they are taken out to sea in a boat, released into the water to do their work, and then jump back into the boat and return home. At least on one occasion animal activists cut the nets to free the dolphins within, and the dolphins wanted no part of it.

I don't think captivity is necessarily a horrendous thing, at least for some dolphins and under the right conditions. They have their individual preferences and inclinations, just as we do. Just as some people are definite dolphin lovers, while others couldn't care less, some dolphins seem genuinely interested in people and enjoy working with us; others come across more like, "Well, it's a living." I only wish we had some way of knowing the difference at the outset.

Some animal-rights activists argue that all captive dolphins should be set free, without regard to how long they've been in captivity or their readiness for life in the wild again. They simply assume the dolphin will be healthier and happier in the ocean. Not necessarily. It could be a bit like taking a person who's lived in the city for twenty years, setting him naked in the woods, and saying, "There, now you're free—be happy."

The idea of releasing dolphins that were born in captivity seems largely untenable. Captive-born dolphins are unlikely to make it in the wild, inexperienced as they are in catching their own food, and in dealing with predators and such natural conditions as tides, water temperature extremes, and currents. They wouldn't know the ropes socially, either. Wild dolphins normally learn these things as young calves from their mothers and other associates. At capture time with Echo and Misha, we specifically wanted dolphins that had been away from their mothers awhile, so they would already have learned these things.

Captive-born animals also could disrupt the wild community into which they are placed. Often the population in a given area is apt to be at or near the maximum of what that environment will hold, so bringing in additional dolphins could put stress on the existing group. The locals are probably also genetically adapted for life in their particular ecosystem, and the new dolphins could weaken their genetic stock. They also could bring in new diseases for which the local community is unprepared and lacks natural immunity.

The need for some dolphins in captivity doesn't necessarily mean we need to capture any more from the wild, however. In fact, there is already a voluntary moratorium on captures of bottlenose dolphins (though not on any other species at this point) in the

United States. Bottlenose dolphins breed well in captivity, and the three hundred–some already in aquariums and research facilities around the country are sufficient to keep the population going and to maintain genetic diversity.

Captive-born dolphins are likely to take more readily to life with humans than wild-caught ones, anyway. That also would mean no further disruptions of the wild communities. In a sense we can have our cake and eat it, too, by maintaining a population of captive-born dolphins; it would give people the opportunity to see dolphins in captivity without taking any more of them from the wild.

Randy points out that the issue isn't just the number of dolphins captured from the wild but the fact that each individual has a role to play, and the loss of one dolphin can have serious repercussions for the rest of the community. His recent findings regarding survivorship and reproductive success indicate that calves born to females in larger and more stable groups are more likely to survive. The females in the band tend to work together for mutual protection of the calves. If even one female is captured out of the group, it could affect the remaining females' chances for raising their offspring successfully. The capture of just one male can also have profound effects, particularly if he's part of a pair bond; his buddy will be unlikely to find a replacement and will face a greatly increased chance of an early death. Capturing dolphins from the wild doesn't just deplete the number of animals in an area, then, but can disrupt the social structure of an entire community.

Some may object to the notion of developing a captive-bred strain—making "dogs" of dolphins. I've had some qualms on that matter myself; after a few generations captive-born animals may not be quite the same creatures their wild cousins are, particularly if people should start intentionally breeding them for specific traits. I confess I'm not eager to see the Pekinese equivalent of a dolphin. We can't rightfully object to that, however, without arguing similarly that all domestication of animals is wrong—dogs, cats, horses, and sheep should all be set free, by that logic. Just because dogs aren't wolves doesn't make them any less interesting or valuable. Besides, we still have wolves as well as dogs. I think there may be room in the order of things for "domestic" dolphins as well as wild ones—especially if having some "domestic" ones around helps make us more aware of the wild ones and more willing to work to protect both them and their habitats.

A philosophy professor I know wrote a book chapter arguing that dolphins might be considered "persons." Philosophically speaking, the term "human being" (or *Homo sapiens*) refers to our particular species, but the term "person" could be applied more broadly to any being that demonstrated certain traits and therefore warranted certain rights. Much as I think dolphins are extraordinary creatures, I'm wary of raising them above the rest of the animal kingdom, of saying that dolphins—along with humans, of course—outshine everyone else. Thus far, anyway, we have more concrete evidence of the intelligence, language abilities, tool use, culture, and so on for chimps than we do for dolphins. They are certainly much closer to us genetically. Are we to consider them persons, too? Maybe. So where do we draw the line between "personhood" and "nonpersonhood?"

I think the point isn't to elevate dolphins (or the species of your choice) to our level, but rather to see ourselves with greater humility. I don't believe we are sole claimants of such traits as intelligence, consciousness, or even soul. I have tried to portray Echo and

Misha, and dolphins more generally, as real-life flesh-and-blood animals, as mere mortals rather than the angels they are sometimes made out to be, but with a spark of the divine nonetheless. That's true not just of dolphins but of all life on Earth. We are all of us mortal, all of us divine.

Acknowledgments

This book is not only about dolphins but also about the many people from whom and with whom I've been learning about dolphins. I've mentioned relatively few of them by name in the book itself, for fear of inundating the reader. My apologies to those who have been left out. The story comes complete with many of the difficulties, disagreements, misunderstandings, and mistakes that are always part of the picture when dealing with real, live people. I hope it's clear, however, that I have enormous respect and affection for all the people involved. (I hope, too, they may forgive anything herein they might find to their disliking.) There would be no book without them. And while I have drawn upon the memories and the thinking of many others in writing this book, the viewpoint—and the responsibility for any errors—is my own.

There would, of course, be no book without the dolphins Echo and Misha; they are in my heart always. I am also profoundly indebted to Josephine, Arrow, the Lags, and the dolphins of Tampa Bay and Sarasota Bay.

I can't begin to express the extent of my appreciation and gratitude to Ken Norris; none of this could have happened without him. Likewise, neither the Echo and Misha project, nor this book, would have been possible without the enormous contributions of Randy Wells, Michelle Jeffries Wells, Kim Bassos Hull, and Howard Rhinehart. Beaucoup thanks, too, to Casey Silvey, who has helped so very much in so many ways. Other dolphin crew members who not only helped care for Echo and Misha but whose recollections and support I relied upon in writing this book include Ania Driscoll-Lind, Dawn Goley, Jill Madden, Holly Sargeant-Green, Kim Urian, and Paula Wolfe.

Much gratitude, also, to "dolphinettes" Chris Ames, Kirstin Anderson, Connie Arthur, Barbara Brunnick, Joan (Lajala) Carroll, Sandy Davis, Karen Hart, Meg Hudson, Nina Morris, Gene Nomicos, Kim Pfeiffer, Kelli Rasmussen, Lisa Takaki, Marcie Tarvid, Bill Thurman, Danielle Waples, and Jackie Weiner.

Many thanks as well to Drs. Dave Casper, Jay Sweeney, and Forrest Townsend for the excellent veterinary care they provided; to Ken Marten, the project's bioacoustician, and to many others of the staff and students of Long Marine Lab, especially Ted Cranford and Carl Schilt; to John Sigurdson and others at NOSC in Hawaii for their assistance in data analysis; and to student interns Kim Hayes and Alyce Santoro.

Others who assisted in the capture and/or release and follow-up include Rob Barrel, Bill Carr, Jack DeMetro, Brett and Pam Fortenberry, Larry Fulford, Cathy Gilligan, Flip Nicklin, Earl Patterson, Suzanne Pfeiffer, Andy Read, Michael Scott, Lee Sehorne, Barbara Steinicke, Jeff Stover, and James Thorson, as well as many members of the staff at Mote Marine Lab, particularly Jay Gorzelany, Sue Hofmann, and Pete Hull.

Scholarship and grant support from Phi Beta Kappa of Northern California and the Dr. Earl H. Myers and Ethel M. Myers

Oceanographic and Marine Biology Trust allowed me to carry out the echolocation research and analysis.

The members of the Whoo Ha writing group—Robin Drury, Ellen Farmer, Yael Lachman, Mary Patvaldnieks, Sarah Rabkin, Jude Todd, and Jesse Virago—provided much-needed commentary, reassurance, and encouragement. I couldn't have made it, either, without Roberta Friedman's generosity and faith in me and Judy Mathewson's timely assistance.

I'm grateful to all three of my editors at Bantam—Barbara Alpert, Janna Silverstein, and Jennifer Steinbach—for their valuable input, and to Ian and Betty Ballantine for their kindly interest in my book. The efforts of my agent, Mike Hamilburg, and his assistant, Joanie Socola, are greatly appreciated as well.

Last but always foremost, I owe much more than gratitude to Joel Shurkin—*ani l'dodi v'dodi li*—and to Hannah Shurkin Howard, who has changed my life forever.

We hope to keep track of Echo and Misha for the rest of their lives. If you would like to help make this possible, please send your tax-deductible contribution to:

Echo and Misha Fund
c/o Mote Marine Laboratory
1600 Ken Thompson Parkway
Sarasota, Florida 34236

Index

300 / Index

JOEL SHURKIN

ABOUT THE AUTHOR

Carol J. Howard studied psychology, speech pathology, and science writing before beginning work on her Ph.D. in biology. Her writing has appeared in *Psychology Today, Reader's Digest,* and the *San Jose Mercury News,* among other places. She lives with her husband and daughter in Santa Cruz, California.